The Law and Practice
Relating to Pollution Control in Ireland

There are nine other titles in this series:

The Law and Practice Relating to Pollution Control In

Belgium and Luxembourg
Denmark
France
Federal Republic of Germany
Greece
Italy
The Netherlands
The United Kingdom

The Law and Practice Relating to Pollution Control in the Member States of the European Communities: A Comparative Survey

The series will be updated at regular intervals. For further information, complete the enclosed postcard and send it to:
Graham & Trotman Limited
Sterling House
66 Wilton Road
London SW1V 1DE

All the titles in the series were prepared by

Environmental Resources Limited
79 Baker St, London W1M 1AJ (Tel. 01-486 8277; Tx. 296359 ERL G)

for

The Commission of the European Communities,
Directorate-General Environment, Consumer Protection
and Nuclear Safety, Brussels

The Law and Practice
Relating to Pollution Control in Ireland

There are nine other titles in this series:

The Law and Practice Relating to Pollution Control In

  Belgium and Luxembourg
  Denmark
  France
  Federal Republic of Germany
  Greece
  Italy
  The Netherlands
  The United Kingdom

The Law and Practice Relating to Pollution Control in the Member States of the European Communities: A Comparative Survey

The series will be updated at regular intervals. For further information, complete the enclosed postcard and send it to:
Graham & Trotman Limited
Sterling House
66 Wilton Road
London SW1V 1DE

All the titles in the series were prepared by

Environmental Resources Limited
79 Baker St, London W1M 1AJ (Tel. 01-486 8277; Tx. 296359 ERL G)

for

The Commission of the European Communities,
Directorate-General Environment, Consumer Protection
and Nuclear Safety, Brussels

The Law and Practice Relating to Pollution Control in Ireland

Second Edition

Prepared by

Yvonne Scannell M.A.,LL.B. (Cantab.), Barrister
Lecturer in Law, Trinity College, Dublin

for

Environmental Resources Limited

Published by
Graham & Trotman
for
The Commission of the European Communities

Published in 1982 by

Graham & Trotman Limited
Sterling House
66 Wilton Road
London SW1V 1DE

for

The Commission of the European Communities,
Directorate-General Information Market and Innovation,
Luxembourg

EUR 7737

© ECSC, EEC, EAEC, Brussels and Luxembourg, 1982

British Library Cataloguing in Publication Data

Scannell, Yvonne
  The law and practice relating to pollution
  control in Ireland.—2nd ed.
  1. Pollution—Law and legislation—Ireland
  I. Title   II. Environmental Resources
  Ltd.   III. Commission of the European
  Communities
  344.1704'463   KDK910

ISBN 0-86010-313-7 ✓

The views expressed in this publication are those of the author, and should
not be taken as reflecting the opinion of the Commission of the European
Communities.

LEGAL NOTICE

Neither the Commission of the European Communities nor any person acting
on behalf of the Commission is responsible for the use which might be made
of the following information.

Printed in Great Britain by
Robert Hartnoll Limited, Bodmin, Cornwall

The Law and Practice
Relating to Pollution Control in Ireland

There are nine other titles in this series:

The Law and Practice Relating to Pollution Control In

Belgium and Luxembourg
Denmark
France
Federal Republic of Germany
Greece
Italy
The Netherlands
The United Kingdom

The Law and Practice Relating to Pollution Control in the Member States of the European Communities: A Comparative Survey

The series will be updated at regular intervals. For further information, complete the enclosed postcard and send it to:
Graham & Trotman Limited
Sterling House
66 Wilton Road
London SW1V 1DE

All the titles in the series were prepared by

Environmental Resources Limited
79 Baker St, London W1M 1AJ (Tel. 01-486 8277; Tx. 296359 ERL G)

for

The Commission of the European Communities,
Directorate-General Environment, Consumer Protection
and Nuclear Safety, Brussels

The Law and Practice Relating to Pollution Control in Ireland

Second Edition

Prepared by

Yvonne Scannell M.A.,LL.B. (Cantab.), Barrister
Lecturer in Law, Trinity College, Dublin

for

Environmental Resources Limited

Published by
Graham & Trotman
for
The Commission of the European Communities

Published in 1982 by

Graham & Trotman Limited
Sterling House
66 Wilton Road
London SW1V 1DE

for

The Commission of the European Communities,
Directorate-General Information Market and Innovation,
Luxembourg

EUR 7737

© ECSC, EEC, EAEC, Brussels and Luxembourg, 1982

British Library Cataloguing in Publication Data

Scannell, Yvonne
  The law and practice relating to pollution
  control in Ireland.—2nd ed.
  1. Pollution—Law and legislation—Ireland
  I. Title   II. Environmental Resources
  Ltd.   III. Commission of the European
  Communities
  344.1704'463   KDK910

ISBN 0-86010-313-7

The views expressed in this publication are those of the author, and should
not be taken as reflecting the opinion of the Commission of the European
Communities.

LEGAL NOTICE

Printed in Great Britain by
Robert Hartnoll Limited, Bodmin, Cornwall

Summary List of Contents

Preface

This volume is part of a series prepared in the performance of a contract between the Commission of the European Communities and Environmental Resources Limited (ERL). ERL is a consulting organisation specialising in environmental research, planning and management.

In 1976 a first series was published covering the, then, nine members of the Community. The purpose of those volumes was to explain the law and practice of pollution control in each of the Member States and to provide a summary comparing all the countries in a separate comparative volume.

Since that time many changes in legislation arising from both national and Community-wide initiatives have occurred. ERL was therefore asked to prepare a new series providing an up-to-date review of the law and practice relating to pollution control in the Member States of the European Community.

The series comprises nine volumes concerning the law and practice in the Member States:

Belgium and Luxembourg   Ireland
Denmark   Italy
France   The Netherlands
The Federal Republic of Germany   The United Kingdom
Greece

and a summary comparative volume.

The aim of this new series, as in the first, is to provide a concise but fully referenced summary of the letter of the law, and a discussion of its implementation and enforcement in practice. Proposals for new legislation which has been drafted but not yet passed are outlined. Where laws have been introduced to comply with Community-wide requirements this is noted.

The publication has two principal objectives:

to enable the reader to study in outline the provisions in any one Member State; and

to enable a direct comparison between different Member States.

To facilitate comparison between the national reports, each is indexed following a standard format (the Classified Index) to enable easy reference to the relevant sections of each report.

Presenting a nation's laws accurately in summary form is always a difficult task. There is a danger that, out of context, they may be misunderstood. We have therefore tried to give, in the first section of each report, some of the constitutional, legal and administrative background.

A further danger lies in translation. Although in the English texts we have tried to prepare as accurate a translation as possible, only the authors' original texts in their native languages carry their full authority. These texts are also being published in the individual Member States.

The statement of law in each volume is correct to at least 30 June 1981; in some cases more recent revisions have been included during the period of preparation for publication.

The series will be updated at regular intervals; to receive further details readers should complete the enclosed postcard and send it to the publisher.

ERL would like to acknowledge and express its thanks for the contributions from the national authors and for their cooperation in the preparation of the series.

Finally, ERL also acknowledges the assistance provided by many agencies, which have freely given information and advice, and the help and guidance given by Monsieur Claude Pleinevaux, Mr Grant Lawrence and other members of the Directorate of Environment, Consumer Protection and Nuclear Safety of the Commission of the European Communities.

1982   Environmental Resources Limited
                                                          London

Detailed List of Contents

CONTENTS

Table of Statutes*

*References in this Table are to section numbers.

Table of Statutory Instruments, Rules and Orders*

*References in this Table are to section numbers.

Table of Cases*

*References in this Table are to section numbers.

Foreword

Contrary to the expectations of those who saw environmental concern as a temporary elitist phenomenon soon to be dissipated by the harsh realities of reconciling changes in material standards and unemployment ratios, the law relating to pollution control in Ireland has developed at an unprecedented rate in the last decade. The legal framework for the control of pollution has been considerably reformed and the powers of public authorities to set, control, monitor and enforce standards have been significantly increased. The law governing this subject is still drawn from a number of diverse sources and the aim of this book is to bring together the more important legal materials relevant to pollution control and, where possible, to describe the practical operation of pollution control laws.

My best thanks are due to Professor B. McMahon of University College, Cork, for stimulating my interest in environmental law, to Dr J. Phillips of Durham University for reading the typescript, to Bridget Ivory, Orla Sheehan and Margot Aspell for typing the manuscript and to Sean O h-Eigeartaigh for his advice and support.

October 1980   Yvonne Scannell
                                                        Trinity College, Dublin

The law and practice herein described is as of 1 April 1981. The opportunity was taken, however, at proof stage to incorporate major legal developments from that date to 1 May 1982. Descriptions of these developments have not been annotated and are necessarily brief.

1
The Constitution, Public Authorities and Interest Groups

1.1 TYPE OF NATIONAL CONSTITUTION

1.1.1 The Legislature

The 'sole and exclusive power' of making laws for the State lies with the Oireachtas which consists of the President and two Houses, viz. a House of Representatives called the Dail and a Senate.[1] The Oireachtas is empowered to delegate powers to make administrative rules and regulations to subordinate bodies or departments.[2] Accordingly, a vast amount of delegated legislation made under provisions in various enabling statutes emanates from designated ministers, departments and subordinate bodies. The Oireachtas is prohibited from enacting any law which is repugnant to the Constitution or to any provision thereof and any law so declared by the High or Supreme Court is invalid to the extent of its repugnancy.[3] In practice, the Government, backed by a majority in the Dail, enjoys a monopoly of initiative in proposing legislation and other motions. The Constitution was amended by referendum in 1972 to enable Ireland to become a member of the European Communities. Under section 2 of the European Communities Act 1972, the treaties governing the European Communities and existing and future Acts adopted by Community institutions become part of Irish domestic law. Section 3 of that Act enables a Minister of State to make regulations for enabling section 2 to have full effect. A number of EEC directives on pollution control have been implemented by ministerial regulations made under section 3.[4]

1

1.1.2 The Executive

The executive power of the State is exercised by or on the authority of the Government.[5] Executive action may be questioned and criticised by means of parliamentary questions and by debates on the annual estimates.

1.1.3 The Courts

1.1.3.1 CRIMINAL JURISDICTION

The Constitution provides that justice must be administered in courts established by law.[6] Criminal jurisdiction relevant to pollution control is exercisable by the District Court, the Circuit Court, the Central Criminal Court, the Court of Criminal Appeal and the Supreme Court.

1.1.3.1.1 The District Court

The criminal jurisdiction of the District Court includes trying the following offences:

(i) Minor offences created by statute and stated to be triable summarily. The maximum penalty for these offences does not normally exceed £600 and/or 6 months' imprisonment. All offences created by regulation under section 3 of the European Communities Act 1972 are triable in the District Court as the Act prohibits the creation of indictable offences.[7]

(ii) Indictable offences specified in the schedule to the Criminal Justice Act 1951, as amended, if:
(a) the Court is of the opinion that the facts alleged or proved constitute a minor offence, and
(b) the accused, on being informed of his constitutional right to trial by jury,[8] agrees to be tried summarily.

(iii) Indictable offences where the accused pleads guilty and the District Justice is satisfied that he understands the charge provided that the Director of Public Prosecutions does not object to a summary trial.

(iv) Crimes committed by children under sixteen.

In addition a District Justice is empowered to carry out a preliminary investigation on the basis of which he may send an accused forward for trial where an indictable offence not triable summarily is involved.

1.1.3.1.2 The Circuit Court

The Circuit Court has the same criminal jurisdiction as the Central Criminal Court save for treason, murder, piracy, certain offences under the Treason Act 1939, and the Offences Against the State Act 1939.

1.1.3.1.3 The Central Criminal Court

The Central Criminal Court is the High Court exercising its criminal jurisdiction. It has original criminal jurisdiction to try crime on indictment.

1.1.3.1.4 Appellate Criminal Jurisdiction

Appellate criminal jurisdiction is enjoyed by all of the above-mentioned courts except the District Court. The Circuit Court hears District Court appeals. The High Court has jurisdiction to make orders directed to any inferior tribunal for the purpose of reviewing its orders or decisions and to give rulings on questions of law submitted to it by a District Court. The Court of Criminal Appeal hears appeals from the Central Criminal Court and Circuit Courts. Supreme Court jurisdiction includes hearing:

(i)   Appeals from the Court of Criminal Appeal where a point of law of exceptional public importance, on which the opinion of the Supreme Court is desirable, is involved.[9]

(ii)   Appeals from the Central Criminal Court against interlocutory orders.

(iii)   Hearing 'cases stated' under section 16 of the Courts of Justice Act 1947.

(iv)   Appeals from High Court decisions on applications for one of the prerogative orders.

(v)   Appeals from High Court decisions on cases stated from District Courts.

(vi)   Cases stated from Circuit Courts.

1.1.3.2 CIVIL JURISDICTION

Civil jurisdiction relevant to pollution control is exercisable by the District Court, the Circuit Court, the High Court and the Supreme Court. The Courts Act 1981 substantially increased the jurisdictions of the inferior courts.

1.1.3.2.1 The District Court

The District Court has general civil jurisdiction in contract and in tort where the claim does not exceed £5,000 and in proceedings at the suit of the State or of any official thereof where the amount involved does not exceed £5,000. It has unlimited jurisdiction in all licensing matters except for new on-licences and new club licences which are dealt with by the Circuit Court.

1.1.3.2.2 The Circuit Court

The Circuit Court has civil jurisdiction in contract and tort where the amount claimed does not exceed £15,000 and in proceedings at the suit of the State or of any official thereof where the amount involved does not exceed £15,000. Parties before the Circuit Court may extend its jurisdiction by mutual consent. It has exclusive jurisdiction to hear applications for new on-licences and new club licences.

1.1.3.2.3 The High Court

The High Court has unlimited civil jurisdiction. The constitutional validity of any laws may not be questioned in any inferior court.[10] Proceedings by individuals for the enforcement of important provisions in the Local Government (Planning and Development) Act 1976, and the Local Government (Water Pollution) Act 1977, must be initiated in the High Court.[11]

1.1.3.2.4 The Supreme Court

The Supreme Court is the Court of Final Appeal. It hears appeals from High Court decisions and cases stated from the Circuit Court.

1.1.3.2.5 The Court of Justice of the European Communities

Since a great deal of environmental law originates in the European Community, the European Court has, potentially at any rate, a role to play in the enforcement of many pollution controls. The European Commission or another member state may complain to the European Court of Justice if they consider that a member state has violated the Community Treaties.[12] If the complaint is justified, the Court will require the errant member state to fulfil its obligations.[13] Ireland has never been brought before the European Court for failure to implement obligations under the EEC Action Programmes on the Environment for 1973–77 or 1977–81 but preliminary action under article 169 has been taken on a number of occasions to accelerate the implementation of some environmental directives.[14] Proceedings may also be taken for the annulment of regulations, directives and decisions made by the Council or the Commission on the grounds of lack of competence, infringement of an essential procedural requirement, infringement of the Treaty or of any rule of law relating to its application, or misuse of powers.[15] The right to bring an action for annulment is not confined to the Community Institutions or member states but may, in limited circumstances, also be enjoyed by individuals.[16] Sometimes also, interested persons may bring actions against the Council or the Commission where, in violation of the Treaty, they have failed to take action for the benefit of a particular person.[17] National courts or tribunals may refer questions on the interpretation of the treaties and on the validity and the interpretation of acts of Community institutions to the Court of Justice which has exclusive competence to determine these questions.[18]

1.2 SOURCES OF LAW

The law relating to pollution derives from four principal sources:

(i)   The Constitution

(ii)   Common Law

(iii)   Statute Law

(iv)   European Community Law

1.2.1 The Constitution

The Constitution is silent on the express question of whether the citizen is entitled to a clean and healthy environment. However, article 40.3.2 does provide that 'the State shall, in particular, by its laws, protect as best it can from unjust attack and, in the case of injustice done, vindicate the life, person, good name and property rights of every citizen'. In *Ryan* v. *The Attorney General*,[19] it was held that the personal rights mentioned in the aforesaid subsection were not exhausted by the enumeration thereof, and that one of the unmentioned personal rights which the citizen enjoys is the right to 'bodily integrity'. This right was defined in the High Court to mean that 'no mutilation of the body or of any of its members may be carried out by any citizen under authority of law except for the good of the whole body, and that no process which is or may, as a matter of probability, be dangerous or harmful to the life or health of citizens or of any of them may be imposed (in the sense of being made compulsory) by an Act of the Oireachtas'.[20] The Ryan case involved a challenge to the constitutional validity of a statute which, it was alleged, obliged an individual to use water containing an additive (fluoride) hazardous to health. The plaintiff was unsuccessful but had her arguments as to the harmful effects of imbibing fluoride been accepted, the provisions of the statute requiring the addition of this substance to public water supplies would have been declared unconstitutional. In *The State (C.) v. Frawley*,[21] Finlay J. extended the principle in the Ryan case to prevent an act or omission of the Executive (as distinct from the Oireachtas) which 'without justification, would expose the health of a person to risk or danger'. It would not be difficult to argue that the principles in the Ryan and Frawley cases should be extended to the acts of administrative bodies like local authorities. On the basis of these cases, it is possible to argue that there exists under the Constitution a limited right not to be polluted.

1.2.2 Common Law

Although common law remedies for pollution control do exist, in practice there are many obstacles to their effective use for this purpose. Nowadays greater reliance is placed on legislative measures establishing an administrative machinery for the control of pollution, leaving common law remedies as a measure of last resort. The main common law remedies are actions for nuisance, trespass, negligence and under the rule in *Rylands* v. *Fletcher*.

1.2.2.1 NUISANCE

Nuisance may be public or private or both. Public nuisance is both a crime and a civil wrong but to be 'public' the nuisance must materially affect the reasonable comfort and convenience of life of a class of citizens. Proceedings for public nuisance may be brought by the Attorney General but an individual may also bring them if he suffers damage over and above that suffered by the public at large.

Private nuisance usually consists of an 'act of wrongfully causing or allowing the escape of deleterious things into another's land—for example, water, smoke, smelly fumes, gas, noise, heat, vibrations, disease-germs, animals and vegetation'.[22] The essence of private nuisance is an unlawful interference with a person's use or enjoyment of land, or of some right over or in connection with it.[23] Accordingly, the plaintiff must have a proprietary or possessory interest in land which has been interfered with and the action is not available to persons lacking a property connection. Once the existence of a nuisance has been proved, the defendant must prove that the interference is justifiable. It is this aspect of the tort which has given it its advantage over negligence where the plaintiff must prove the defendant's fault. Substantial damage must be proved before a plaintiff can sue in nuisance. The standard for determining whether damage is substantial is objectively judged in the sense that the idiosyncracies of the hypersensitive plaintiff must be discounted unless caused by the nuisance.[24] But the character of the neighbourhood which the alleged nuisance affects may be relevant in determining whether a remedy is available. As Thesiger L. J. said in *Sturges* v. *Bridgman:*[25] 'what would be a nuisance in Belgrave Square would not necessarily be so in Bermondsey'. The remedies available to a successful plaintiff are damages and/or an injunction. In addition, a person affected by a nuisance may, in certain circumstances, abate it.

1.2.2.2 TRESPASS

Pollution may constitute a trespass. A trespass may be committed by, *inter alia*, placing or projecting any material object on land or causing any physical or noxious substance[26] to cross the boundary of a plaintiff's land or even simply to come into physical contact with the land, though there be no crossing of the boundary.[27] Trespass is essentially an injury to a plaintiff's interest in the possession of property. The tort is complete upon a tangible invasion of the plaintiff's property, however slight, whether or not damage results, but potential difficulties in succeeding in trespass actions for pollution lurk in the requirements that a trespass invasion be direct and intentional. Some courts hold that the interference is not direct if an intervening force such as wind or water carry pollutants on to a plaintiff's land.[28] However, it has been argued that the concept of 'direct' injury where pollution is concerned is repudiated by the contemporary science of causation which establishes that 'atmospheric or hydrologic systems assure that pollutants deposited in one place will end up some place else, with no less assurance of causation than the blaster who watched the debris rise from his property and settle on his neighbour's land'.[29] Damages and/or an injunction are the usual remedies.

1.2.2.3 NEGLIGENCE

Negligence consists of a breach of a legal duty of care owed by one person to another as a result of which that other suffers damage.[30] The difficulty with this action from the point of view of a plaintiff who seeks to control or prevent pollution is that it is frequently difficult to prove negligence on the part of the polluter, damage by the pollutant and the

7

foreseeability of that damage. The usual remedies are damages and/or an injunction.

1.2.2.4 THE RULE IN *RYLANDS* v. *FLETCHER*[31]

The rule in *Rylands* v. *Fletcher* imposes strict liability upon the occupier of land who brings and keeps on it anything liable to do damage if it escapes. The rule is qualified by the requirement that the use of land must be 'non-natural'. But non-natural uses include many activities which are simply high in risk: water stored in bulk, electricity in bulk, storage of a motor-vehicle with a tank full of petrol in a garage, the collection of toxic waste in a dump, etc.[32] The action, unlike an action in nuisance, may be brought by anyone who has suffered damage. The remedies are damages and/or an injunction.

1.2.3 Statute Law

Most modern pollution control law originates in statutes and subordinate legislation made thereunder. The common law of nuisance has also been modified by statute and a number of statutory nuisances have been created which embrace various types of pollution.[33] Sanitary authorities are obliged to inspect their districts for such nuisances[34] and on receipt of a complaint about a statutory nuisance to take steps to ensure that it is abated.[35] Prosecutions for statutory nuisances may be brought by persons aggrieved, inhabitants of sanitary districts or sanitary authorities.[36] The tendency in modern pollution control legislation is to extend rights to enforce pollution control to the ordinary citizen[37] thus limiting further the relevance of old common law remedies and ensuring that persons other than public authorities are involved in protecting the environment.

1.2.4 European Community Law

The adoption by the European Council of a Community Action Programme on the Environment in 1973 was probably the most significant factor in motivating Irish authorities to enact, update and extend laws on pollution control. Indeed, were it not for the necessity of complying with various EEC directives, it is doubtful whether many of the pollution controls currently in force would have been introduced in their present form at all. Since 1973 the Community has adopted about sixty legislative texts on pollution control including fifteen on water pollution, ten on air, seven on waste, eight on noise pollution and four on the protection of the environment, land and natural resources.[38]

1.3 RESPONSIBILITIES OF GOVERNMENT DEPARTMENTS AND AGENCIES IN ENVIRONMENTAL POLLUTION MATTERS[39]

1.3.1 Department of the Environment

The Minister for the Environment has overall responsibility for protecting and improving the physical environment. He is advised by the Water Pollution and Environment Councils.[40] The Department of the Environment has two sections with responsibilities for pollution control matters, viz.

(i) *An Environmental Policy Section* which is responsible for the review of environmental legislation; the implementation of the EEC Action Programme on the Environment; the implementation of the environmental programmes of other international organisations; and the review and coordination of local authority standards and practices in environmental matters.

(ii) *A Pollution Control Section* which is concerned with air and water pollution, pollution by improper waste disposal, the implementation of various EEC directives on waste and air pollutants, the control through the alkali inspectorate of industrial emissions to the atmosphere and the coordination of arrangements for the clearance of oil pollution from beaches and immediately offshore.

The Department has produced two documents on environmental matters since 1972—the *Report on Pollution Control (1979)*[41] and Memorandum No. 1 on *Water Quality Guidelines (1979)*.[42] As yet, there is no national policy on environmental protection. While the laws on pollution control have been considerably improved since 1973, there are no or inadequate controls in some areas; monitoring of compliance with existing laws and the enforcement of existing controls by public authorities is virtually non-existent; penalties provided in legislation are relatively

9

small and there has been no real attempt to implement the 'polluter pays' principle.

The Department is the central control authority for local authorities in their various capacities. An Foras Forbartha (the National Institute for Physical Planning and Research) operates under the aegis of the Minister for the Environment.

1.3.2 Department of Energy

This department is responsible for pollution control from mineral petroleum exploration and development, and for enforcement of EEC Directive 75/716/EEC on the sulphur content of certain liquid fuels.[43] The Nuclear Energy Board reports to the Minister for Energy.

1.3.3 Department of Agriculture

This is concerned with the effects of environmental pollution on agriculture, the pollution implications of agricultural practice including the control of agrichemicals, and the administration of the Farm Modernisation Scheme. An Foras Taluntais (the Agricultural Institute) and An Comhairle Oiliuna Talamhaiochta (the Agricultural Training and Advisory Council) report to the Minister for Agriculture.

1.3.4 Department of Fisheries and Forestry

This is responsible for licensing deleterious discharges at sea under the Fisheries (Consolidation) Act 1959. It has powers under the Fishery Harbours Centres Act 1968, for pollution prevention. It enforces water pollution controls under the Fisheries Acts 1959–80 and the Local Government (Water Pollution) Act 1977. Regional Fisheries Boards report to the Minister. This department is also concerned with pollution effects on wild-life and wild-life conservation.

1.3.5 Department of Tourism and Transport

This is responsible for the enforcement of oil pollution controls, the authorisation of dumping at sea, the enforcement of provisions of the Foreshore Act 1933, the adoption and enforcement of noise certification standards in aircraft, and of procedures to reduce aircraft noise in residential areas. Harbour authorities report to the Minister.

1.3.6 Department of Labour

This is responsible for the implementation of the Dangerous Substances Act 1972, and for the protection of workers exposed to pollution.

1.3.7 Department of Health

This is concerned with environmental health problems and food contamination. It has overall supervision of Health Boards who are involved in monitoring water quality.

1.3.8 Department of Defence

This deals with the clearance of oil spillages at sea.

1.3.9 Environment Council

The Environment Council is a non-statutory body set up by the Minister for the Environment to advise him generally on his environmental functions. Its membership is reflective of sectoral as well as environmental interests. The Council has been charged by the Minister with developing an environmental policy—with particular emphasis on the litter problem. To date it has published two documents—*Towards an Environment Policy*[44] and *Litter and the Environment*.[45] A Litter Bill was introduced in March 1981 to implement the Council's proposals and to make other necessary provisions for the control of litter.

1.3.10 Water Pollution Advisory Council

The Water Pollution Advisory Council (WPAC) was appointed in 1977 in accordance with the provisions of section 2 of the Local Government (Water Pollution) Act 1977, to replace an informal council which had been in operation since 1975. It is representative of various interest groups affected by water pollution legislation. Its statutory function is to make recommendations to the Minister for the Environment in relation to his water pollution control functions and responsibilities, either on its own volition or at the request of the Minister. The Minister is obliged to consult the Council before exercising his functions under sections 3(10), 24, 25, 26 and 27 of the Local Government (Water Pollution) Act 1977. The WPAC has proved to be an independently-minded and responsible body genuinely concerned with the development and implementation of water pollution controls.[46]

1.3.11 The Inter-departmental Advisory Committee

The Inter-departmental Advisory Committee is not, strictly speaking, an agency but it does perform functions which might be assumed by a more formal body. Its function is 'to coordinate the various activities of the public sector affecting environmental matters'.[47] By Government decision it has the following terms of reference:

(i) to be a means of communication between Departments in environmental matters, to promote coordination and joint consideration by Departments where appropriate, particularly in relation to EEC and international measures;

(ii) to examine arrangements relating to environmental protection and improvement and to consider their adequacy;

(iii) to provide such input as may be called for to the working of the Environment Council;

(iv) to monitor the consideration of matters arising from the Report on Pollution Control.[48]

1.4 RESPONSIBILITIES OF NATIONAL, REGIONAL AND LOCAL AUTHORITIES IN ENVIRONMENTAL POLLUTION MATTERS

1.4.1 Planning Appeals Board

The Planning Appeals Board was established in 1977 to hear appeals and to deal with references under the Local Government (Planning and Development) Acts 1963–1976. In 1978 responsibility for dealing with appeals under the Local Government (Water Pollution) Act 1977 was assigned to it. Membership of the Board may consist of four to ten ordinary members appointed by the Minister for the Environment and a Chairperson who must be a serving High Court judge or a former holder of judicial office appointed by the Government.[49] The Board has a duty to keep itself informed on the policies and objectives of public bodies whose functions are concerned with proper planning and development.[50] The Minister may give the Board general directives on planning and development policy to which it must have regard but he may not address it on any specific case.[51] The Act provides that all such directives must be published in a prescribed manner.[52] The more important functions of the Board are described in later sections.[53] The Board enjoys a reputation for fairness and impartiality.

1.4.2 Regional Development Organisations

Regional Development Organisations are non-statutory bodies representative of local authorities and of other development interests such as industrial development and tourist authorities. The country has been divided into nine regions for economic development purposes. Regional development officers report on, *inter alia*, physical and infrastructural factors which are important for regional development programmes. Their main concern is the promotion of economic development but they are also concerned with the pollution control implications of industrialisation and they have coordinated arrangements for dealing with oil pollution on a regional basis.

1.4.3 Local Authorities

Local authorities consist of 27 county councils, 4 county borough corporations, 7 borough corporations, 49 urban district councils and 28 boards of town commissioners. County councils and county borough corporations were established under the Local Government (Ireland) Act 1898. Borough corporations were reformed and reconstituted by the Municipal Corporations (Ireland) Act 1840. Town commissioners came into existence in some cases under an Act of 1828, in others by Private Act, but generally under the Towns Improvement (Ireland) Act 1854. The modern urban district council is the successor of the urban sanitary authorities created under the Public Health (Ireland) Act 1878.[54] County councils exercise their jurisdiction over that part of a county unit which is not within the jurisdiction of other local authorities. County borough corporations are responsible for the cities of Dublin, Cork, Waterford and Limerick. Borough corporations, urban district councils and town commissioners have functions over a number of smaller towns.

Members of local authorities are elected, usually every five years, on a system of proportional representation. Elections are fought almost exclusively on a party political basis. All qualified persons registered as local electors are entitled to vote. Elected members have overall responsibility over the activities of their particular local authority and certain exclusive powers (known as 'reserved functions') while administrative and executive matters are left to a County or City Manager and staff.[55]

All local authorities except boards of town commissioners are also planning and sanitary authorities. Planning authorities are responsible for the control of land use in their areas.[56] Sanitary authorities have a wide range of functions including functions relating to air[57] and water pollution control,[58] sewage,[59] water supplies,[60] waste disposal[61] and the suppression of statutory nuisances.[62] The larger local authorities (county councils, borough corporations, and Dun Laoghaire corporation) are responsible for the administration and enforcement of water pollution controls relating to discharges to waters established under the Local Government (Water Pollution) Act 1977.[63]

Local authorities may, and do, cooperate with each other and other statutory bodies. They are empowered to exercise or perform a function, power or duty of another local authority or statutory body.[64] The Minister for the Environment is responsible to the Oireachtas for their activities and his powers include powers of approval in relation to many local authority plans; powers of action in default; powers to require coordination and cooperation between different local authorities; powers to give guidance on policy matters; and the ultimate power to dismiss members of a local authority in special legally defined circumstances. Because of their substantial financial dependence on central government finance, local authorities are, in practice, subordinate to the Department of the Environment.

1.4.4 County Committees of Agriculture

Under the Agriculture Act 1931, as amended,[65] each county council must appoint a committee of agriculture. Each committee has corporate status. Two-thirds of the members are appointed by the county council and the remainder on the nomination of active voluntary rural organisations. County committees of agriculture and the Department of Agriculture are responsible for agricultural advisory and training services and the administration of grant-aids to farmers. While they have no statutory functions in relation to pollution control, in practice, they do influence, and when approving grant aids, require, farmers to adopt good agriculture practices in matters such as the disposal of farm wastes[66] and the proper use of agrichemicals. The advisory and training functions of the county committees were transferred to the An Comhairle Oiliuna Talmhaiochta (ACOT) in July 1980 under the provisions of the National Agricultural Advisory, Educational and Research Authority Act 1977, and the Agriculture (An Comhairle Oiliuna Talmhaiochta) Act 1979.

1.4.5 Fisheries Boards

The Fisheries (Consolidation) Act 1959 provided for the establishment of boards of conservators in the seventeen fishery districts specified in the Act.[67] Their functions included protecting fish life generally and for this purpose they had prosecution powers for water pollution under the 1959 Act and under the Local Government (Water Pollution) Act 1977. In practice, they took their duties with respect to water pollution control far more seriously than most local authorities.[68] They were responsible to the Minister for Fisheries. The Fisheries Act 1980, provided for the dissolution of boards of conservators and for their replacement by a central fisheries authority and seven regional fisheries boards.

15

1.4.6 Harbour Authorities

Harbour authorities were established under the Harbours Acts 1946 and 1947, and are responsible for the administration of harbours legislation in the commercial harbours scheduled in the 1946 Act. They have powers under the Harbours Acts and the Oil Pollution of the Sea Acts 1956–1977 to control oil pollution in harbour areas.[69] They also have responsibilities in relation to the control of dangerous substances under the Dangerous Substances Act 1972 and regulations made thereunder.[70] Harbour authorities are responsible to the Minister for Transport.

1.5 INDEPENDENT ADVISORY BODIES WITH FUNCTIONS RELEVANT TO POLLUTION CONTROL

1.5.1 Institute for Industrial Research and Standards

The Institute for Industrial Research and Standards (IIRS) is a State-sponsored body responsible to the Minister for Industry and Commerce, by whom the Board is appointed. It was established under the Industrial Research and Standards Act 1946 and its functions are further defined in the Industrial Research and Standards Act 1961. Its only statutory function directly relevant to pollution control is prescribed in section 6 of the 1961 Act which requires it to promote or facilitate 'the utilisation of the waste products of industry' but IIRS perceives its role in promoting the utilisation of the natural resources of the State as bringing with it an obligation 'to ensure that in the development of those resources other natural resources, such as air, water and amenities are not damaged'.[71]

IIRS has five main areas of activity in the environmental field.

(i)   Its 'primary' responsibility is 'to help industry to meet the environmental constraints set by local and national authorities'.[72]

(ii)   In the absence of clearly defined statutory or local requirements regarding the environment, it advises Government and local authorities on 'acceptable norms for industrial practice in relation to effluents, emissions, noise and vibration, solid and hazardous wastes'.[73] It also prepares national standard recommendations and codes of practice for pollution control.[74]

(iii) It participates in public service and EEC working groups on aspects of pollution control and is represented on most official and semi-official pollution control committees.[75] In this capacity the IIRS provides the scientific and technical expertise which the administrative Civil Service lacks.

(iv) It acts as a consultant to the Industrial Development Authority (IDA) on environmental matters. Since 1972 it has carried out a limited form of environmental impact assessment on all industrial projects for which IDA grant-aids were sought. Where water pollution is in question, the aid of An Foras Forbartha is enlisted and joint assessments and recommendations are made. The suitability of proposed pollution control measures is assessed and environmental quality and/or emission standards are recommended to the IDA and incorporated as conditions for grant-aids.[76] The IDA sends copies of the recommended standards to the local authority of the area in which the industry is or will be situated and to the Department of Fisheries. In practice, these standards are almost invariably adopted by local authorities and incorporated as conditions for planning permissions and other authorisations.[77]

(v) It acts as a consultant to Government departments and local authorities carrying out their environmental protection functions. It has carried out environmental impact assessments on many major new industries on behalf of local authorities.[78]

Although standards recommended by the IIRS are not in themselves mandatory, they may, and frequently do, become so by incorporation as conditions for grant-aids, planning permissions and authorisations of various kinds necessary for polluting activities. Pollution control authorities have not got the expertise, staff or facilities to set environmental standards themselves and they rely greatly and sometimes exclusively on the IIRS and/or An Foras Forbartha to provide them with technical information and advice. In practice, therefore, the position is that the IIRS is often the real—as distinct from the formal—environmental standard-setter in Ireland. Whether the IIRS should enjoy this role has been questioned. The IIRS depends on private enterprise for a good, and increasing, proportion of its finances. Its clients include both public bodies and private entrepreneurs. In a particular case it may act for both. On the topic of its conflicting functions, the IIRS has claimed that it 'cannot or will not support or encourage industrial development where short-term benefits may lead to long-term damage to human health or the environment',[79] and it also points to its record to date, which is

17

reputed to be a good one. In 1977 the IDA-sponsored *A Survey of Pollution in Ireland* stated that 'Among those IDA-sponsored new industries that have established in Ireland in the last five years, there has been almost without exception, no sign of any environmental pollution nor any indication that stringent standards imposed are not being met'.[80] Even if this statement is entirely accurate—and none but the IIRS is competent to comment on this—a system which permits the IIRS to work for both the polluter and the pollution control authorities and thereby constitutes it the de facto and unelected arbiter of environmental standards must surely be lacking.

1.5.2 An Foras Forbartha

An Foras Forbartha (The National Institute for Physical Planning and Research) is a state-sponsored body established in 1964 by the then Minister for Local Government to provide advice and undertake research in physical planning and development, in building and construction, and in roads and water resources. It is responsible to, and works very closely with, the Minister for the Environment and derives the vast bulk of its income from state grants and charges for services. AFF operates a Conservation and Amenity Advisory Service to local authorities on environmental issues, including pollution control. It also collects, collates and processes data on the quality of water resources throughout the country and has established hydrometric teams to monitor water quality in eight regional locations. It assists in the operation of a laboratory system for monitoring water quality in rivers and lakes and the quality of effluents discharged to these bodies. It is also carrying out research into the assimilative capacity of estuaries and coastal regions and the effects of industrial development on the country's water resources. At present, AFF is preparing water quality management plans[81] for local authorities. AFF participates in public service and EEC working groups on aspects of pollution control. With the IIRS, it carries out assessments on the water pollution implications of industries which have applied for IDA grant-aids and both bodies formulate joint recommendations on environmental standards which ought to be observed.[82] The functions of the IIRS and AFF appear to overlap but the institutes themselves have made administrative agreements to prevent duplication of work.[83] In general, the greater proportion of IIRS work is for the private sector while the work of AFF is public sector orientated. AFF has produced more than 150 research publications and technical reports on environmental matters. It sees its role as that of 'an environmental watchdog, free of commitment to any polluting interest'.[84]

1.5.3 Industrial Development Authority

The Industrial Development Authority is a State-sponsored body charged with the promotion of industrial development in Ireland. Under the Industrial Development Act 1969, it has wide statutory powers to give financial and other assistance to qualifying industries. In 1979 grant payments exceeded 79m.[85] Most new and expanding industries receive some kind of IDA economic aid. The IDA requires that industries liable to cause pollution problems be subjected to a limited form of environmental impact assessment by its consultants the IIRS and AFF (for water pollution). Environmental standards recommended by the IIRS and AFF are incorporated as conditions for grant-aids and grants are also made conditional on compliance with conditions attached to planning permissions and the obtaining of any necessary licences. Non-compliance with a term or condition attached to a grant or other payment may result in the grant or payment becoming repayable to the IDA and, in default of being repaid, it is recoverable as a simple contract debt.[86] It is understood that a grant has never been repaid because of non-compliance with pollution controlling conditions. The IDA also builds advance factories and purchases sites for potential industrial development. It has thus developed as one of the greatest determinants of industrial location in the country.

The IDA claims to have a responsible attitude towards the environment. Since 1972 it has taken environmental considerations into account when considering applications for grant-aids. It has been noted in a preceding section that, in practice, environmental standards recommended by IDA consultants are frequently adopted as mandatory standards by pollution control authorities.[87] In 1977 the IDA commissioned IIRS to carry out a national survey of air and water pollution 'to establish that the creation of new industrial jobs in Ireland is not achieved at the cost of polluting our water and air'.[88] This survey is the most comprehensive and detailed analysis of pollution control carried out in Ireland to date. An edited version was published in 1977.[89]

To some extent the IDA involvement in the setting of environmental standards may be justified. It has the practical advantage that it gives the industrialist a good idea of what standards he may be required to meet before he commits himself irrevocably to a project. The absence of mandatory environmental quality and emission standards for many pollutants renders IDA practice even more justifiable if not defensible in principle. The IDA claims with some justification that if it had not established its environmental impact procedures in 1972 and required the installation of pollution control equipment, 'it is unlikely . . . that much pollution control equipment would have been installed in many

19

plants'.[90] This was probably true in the early seventies when pollution control laws were often inadequate or ignored by local authorities.

1.5.4 An Foras Taluntais

An Forus Taluntais (the Agricultural Institute) was established under the Agriculture (An Foras Taluntais) Act 1958. Its functions include reviewing, facilitating, encouraging, assisting, coordinating, promoting and undertaking agricultural research in Ireland. It is responsible to the Minister for Agriculture, to whom it submits its annual report. It has no specific statutory function in the field of pollution control, but its work contributes to the control of pollution, especially agricultural pollution. It is to the agricultural sector what IIRS and AFF are to the industrial sector. It is specifically obliged to disseminate the results of agricultural research, particularly to those engaged in agricultural advisory work, and it has a statutory obligation to advise the Minister for Agriculture on any matter relating to agricultural research or agricultural science on which advice is requested by him. A booklet on *Methods of Treatment of Milk Processing Wastes* published by the Institute is widely used in farming circles. Special sections in the Institute are concerned with pollution, viz. the farm structures and environment department, the soil fertility and chemistry department, the pesticide residue and analytical services unit, soil physics, field investigations, soil biology and animal husbandry departments, the dairy microbiology department and the analytical services laboratory. The Institute has given high priority to pollution matters and has been involved with a number of local authorities in relation to planning applications for large industrial units where it was thought that pollution from the proposed industry could have an unfavourable effect on agriculture in the vicinity.

The Agricultural Institute is comparatively well equipped, well financed and well staffed. It is not subject to the same conflicting interests as are, for example, the IDA and IIRS. Its role as a pollution control authority should not be under-estimated as controls over pollution caused by agricultural activities are largely extra-legal.

1.5.5 National Board for Science and Technology

The National Board for Science and Technology (NBST) was established under the National Board for Science and Technology Act 1977. The Board consists of a chairperson appointed by the Government and up to ten ordinary members who do not represent specific interest groups. Its main functions include advising the Government or the Minister to whom it is responsible on policy for science and technology and related matters; promoting research in the fields of science and technology; and promoting the development of national resources through the application of science and technology.[91] NBST has set up an Environmental Sciences Advisory Group which promotes the development and implementation of a nationally coordinated environmental sciences research programme. NBST and the EEC have funded a number of research projects on the environment as part of the EEC Second Environmental Research Programme (1976–81). The main contribution of NBST to environmental policy-making has been through the publication of the proceedings of various seminars on environmental problems of particular significance in the Irish context and the identification of research projects in areas of national importance.

1.6 SPECIAL INTEREST GROUPS

1.6.1 An Taisce

An Taisce—the National Trust for Ireland—is a voluntary non-profit-making company financed by private donations and subscriptions from members. It is concerned with protecting environmental quality in general and its principal aim is to protect the nation's physical heritage. It is a prescribed authority for the purposes of section 21(1)(a) of the Local Government (Planning and Development) Act 1963. Planning authorities are obliged to send it copies of draft development plans and of draft amendments or variations thereto as well as copies of all planning applications for development in areas of special amenity whether or not a special amenity area order has been made for the area or for developments which would obstruct any view or prospect of special amenity value or special interest.[92] An Taisce has no other specific rights under pollution control legislation but it avails of rights available to the individual.[93] It may be considered as the most active and the most effective environmental watch-dog in the country.

1.6.2 Bord Failte Eireann

Bord Failte Eireann (the National Tourist Board) is a corporate body established under statute and financed by annual Government grant. It

21

is concerned with all aspects of tourist development including the development of amenities. It is a prescribed body under section 21(1)(*a*) of the Local Government (Planning and Development) Act 1963, and planning authorities are obliged to send it copies of draft development plans or variations or amendments thereto as well as copies of planning applications for developments in areas of special amenity or for developments which would obstruct any view or prospect of special amenity value or special interest.[94] Bord Failte encourages community participation in amenity projects by way of competitions such as Tidy Towns, National Gardens and Civic Award Schemes.

1.6.3 Electricity Supply Board

The Electricity Supply Board (ESB) is a statutory corporation which has a virtual monopoly for supplying electricity in Ireland. It owns extensive fisheries in its own right and has powers of control over these. In addition, section 42 of the Electricity Supply (Amendment) Act 1945 gives the ESB a measure of control over rivers and streams serving electricity generating stations. This section prohibits any person, without the written permission of the Board, from discharging or allowing into a river (or into any tributary of any river, or any watercourse connected with any river) which is to be used by the board in connection with the generation of electricity, any chemical or other substance which might injure any part of the generating station or any works subsidiary to it or connected with it. Contravention of the provision is punishable by a £50 fine plus £20 for every day on which the offence is continued. Prosecutions have been taken in a number of cases. The ESB spends considerable amounts in the preservation and development of amenity resources under its control and on fisheries. It employs an administrative officer in its head office and pollution investigation officers in the field. Its fisheries protection staff are instructed to report on cases of pollution which come to their notice in their daily rounds. It has its own central laboratory for investigating, *inter alia*, pollution, together with facilities for BOD, pH and ammonia analyses. When necessary, it avails of the expertise of the public analyst and other bodies, for example, AFF, IIRS and An Foras Taluntais. It has regular consultations with local authorities in their various capacities and is represented on various planning committees and regional development organisations. It operates a close liaison with fisheries interests, especially the Department of Fisheries. It monitors air pollution in a number of locations, and has its own internal standards for emissions.

1.6.4 Others

Other associations concerned directly or indirectly with pollution control include the Waterways Association of Ireland, the Maritime Institute, the Heritage Trust, local field clubs, anglers' and residents' associations and local development committees.

Notes

1. *Constitution of Ireland*, art. 15.1.
2. *Ibid.*, art. 15.2.
3. *Ibid.*, art. 15.3.
4. See Chapter 8.
5. *Constitution of Ireland*, art. 28.2.
6. *Ibid.*, art. 34.1.
7. European Communities Act 1972, s.3(3).
8. *Constitution of Ireland*, art. 38.5.
9. Courts of Justice Act 1924, s.29.
10. *Constitution of Ireland*, art. 34.3.
11. See Local Government (Planning and Development) Act 1976, s.27 and Local Government (Water Pollution) Act 1977, s.11.
12. Treaty Establishing the European Economic Community, arts. 169, 170.
13. *Ibid.*, art. 171.
14. For example EEC Directives 75/442 on waste; 78/178 on waste from the titanium dioxide industry; 76/403 on the disposal of polychlorinated biphenyls and polychlorinated terphenyls.
15. Treaty Establishing the European Economic Community, art. 173.
16. *Ibid.*
17. *Ibid.*, art. 175.
18. *Ibid.*, art. 177.
19. *Ryan* v. *The Attorney General* [1965] I.R. 344.
20. *Ibid.*, p. 345.
21. *The State (C.)* v. *Frawley* [1976] I.R. 365.
22. Salmond, *The Law of Torts* (16th edn), pp 41–42.
23. *Winfield and Jalowicz on Tort* (10th edn), p. 318.
24. Salmond, *The Law of Torts*.
25. *Sturges* v. *Bridgman* (1879) 11 Ch.D. 852, 865. See also *Pembroke* (Earl of) v. *Warren* [1896]1 I.R. 76.
26. *McDonald* v. *Associated Fuels* (1954) D.L.R. 775.
27. Salmond, *The Law of Torts*, p. 38.
28. *Esso Petroleum Co. Ltd* v. *Southport Corporation* [1956] A.C. 218.
29. Rodgers, W. H., *Environmental Law* (1977), p. 156.
30. *Donoghue* v. *Stevenson* [1932] A.C. 562.
31. *Rylands* v. *Fletcher* (1868) L.R. 3 H.L. 330.
32. Salmond, *The Law of Torts*, p. 323.
33. See 3.1 and 4.1.
34. Public Health (Ireland) Act 1878, s.108.
35. *Ibid.*, ss.109–112, 113, 123.
36. *Ibid.*, s.121.
37. See Local Government (Planning and Development) Act 1976, s.27; Local Government (Water Pollution) Act 1977, s.11. See also 2.4.8 and 4.4.7.

38. This was the position in May 1980.
39. See *Report on Pollution Control* (1979), Prl. 6970, GPO, Appendix I.
40. See 1.3.10 and 1.3.11.
41. *Report on Pollution Control* (1979), *supra*.
42. *Memorandum No. 1 on Water Quality Guidelines* (1979), GPO.
43. See 3.6.
44. *Towards an Environment Policy* (1979), Prl. 8215.
45. *Litter and the Environment* (1980), Prl. 8953.
46. See *Annual Reports of the Water Pollution Advisory Council* 1977–1980.
47. Information from the Department of the Environment.
48. *Ibid.*
49. Local Government (Planning and Development) Act 1976, ss. 3, 4.
50. *Ibid.*, s.5.
51. *Ibid.*, s.6.
52. *Ibid.*
53. See especially 2.4.5 and 4.4.3.
54. *Local Government Re-organisation*, Prl. 1572, GPO.
55. See County Management Act 1940, and City and County Management (Amendment) Act 1955.
56. See Chapter 2.
57. See 3.1, 3.2.3, 3.3, 3.10.
58. See 4.1, 4.5, 4.6, 4.9.
59. See Chapter 7.
60. See 4.5.
61. See 8.
62. See 3.1, 4.1, 7.1, 7.2.1, 8.3.6, 8.3.8.
63. See 4.4
64. Local Government Act 1955, s.59.
65. Agriculture (Amendment) Acts 1944, 1958, 1964, 1967, 1973, 1977, 1979.
66. See 4.8.
67. Fisheries (Consolidation) Act 1959, Second Schedule.
68. See 4.2.
69. See 6.
70. See 6.8.
71. *The Role of the IIRS in the Environmental Field* (1977), p. 3.
72. *Ibid.*, p. 4.
73. *Ibid.*
74. *Ibid.*, p. 9.
75. *Ibid.*, Appendix 3.
76. See 1.5.3.
77. See Convery, F. G., 'Some Environmental Policies—Review and Outlook', *Irish Economic Policy*, ESRI, pp. 375–379.
78. *The Role of the IIRS in the Environmental Field*, p. 13.
79. *Ibid.*, p. 4.
80. *A Survey of Pollution in Ireland* (1977), IDA, p. 3.
81. See 4.4.5.
82. See 1.5.1.
83. Memorandum of Understanding between An Foras Forbartha and the Institute for Industrial Research and Standards.
84. *Ibid.*
85. Annual Report of the Industrial Development Authority 1979, p. 87.
86. Industrial Development Act 1969, s.45.
87. See 1.5.1.
88. *A Survey of Pollution in Ireland* (1977), IDA.
89. *Ibid.*

90. *Ibid.*, p. 1.
91. National Board for Science and Technology Act 1977, s.4.
92. See 2.1.2, 2.1.4 and Local Government (Planning and Development) Regulations 1977, art. 25.
93. See 2.4.9 and 4.11.
94. See 2.1.2, 2.1.4 and Local Government (Planning and Development) Regulations 1977, art. 25.

2

Development Control under the Local Government (Planning and Development) Acts 1963 and 1976

Any description of pollution control in Ireland must centre around the provisions and operation of the Local Government (Planning and Development) Acts 1963–1976. These Acts, though not designed for, or particularly suitable as, instruments for the control of pollution, have provided the principal mechanisms for this purpose in the last fifteen years. Their emergence as such is to a large extent due to the absence of any other mechanisms or, more accurately, any other effective mechanisms, for pollution control in an era of rapidly increasing urbanisation and industrialisation.

In recent years, however, the deficiencies inherent in these Acts as techniques for controlling pollution have become increasingly obvious, and modern policy appears to favour the enactment of separate legislation for the protection of specific environmental media (water, air, land) or for the control of environmentally harmful activities (e.g. waste disposal, noise generation).

The Planning Acts operate on two levels: on one they provide for the making and implementation of schemes regulating land use in a general way; on the other they prohibit the development of land unless it is authorised and carried out under, and in accordance with, the permission of the appropriate planning authority.

The Acts are administered at local level by eighty-seven local planning authorities.[1] Appeals from decisions by local planning authorities generally lie with the Planning Appeals Board[2] and thence, on a point of law, to the High Court.

2.1 DEVELOPMENT PLANS

2.1.1 Nature and Content

Statutory land-use plans in Ireland are, with minor exceptions, prepared and implemented by local planning authorities in accordance with the provisions of the Planning Acts 1963–76 and the Local Government (Planning and Development) Regulations 1977.[3] The 1963 Act requires every planning authority to 'make a plan indicating the development objectives for their area'.[4] A development plan must consist of a written statement and a plan which is essentially a map, indicating the development objectives for the area in question.[5] There is no exhaustive definition of the term 'development objectives' but the term does include objectives for physical, economic and social development. Mandatory and permissible objectives are set out in the 1963 Act and in the Third Schedule thereto.[6] Mandatory objectives differ for urban and rural areas. Those for county boroughs, boroughs, urban districts and towns scheduled in the First Schedule to the 1963 Act are objectives:

(i)   for the use solely or primarily (as may be indicated in the development plan) of particular areas for particular purposes, whether residential, commercial, industrial, agricultural, or otherwise:

(ii)   for securing the greater convenience and safety of road users and pedestrians by the provision of parking places or road improvements or otherwise;

(iii)   for development and renewal of obsolete areas;

(iv)   for preserving, improving and extending amenities.

Mandatory objectives for other areas are:

(i)   for the development and renewal of obsolete areas;

(ii)   for preserving, improving and extending amenities;

(iii)   for the provision of new water supplies and sewage services and the extension of existing such supplies and services.[7]

A wide range of permissible objectives for all areas is listed in the Third Schedule under the headings of Roads and Traffic, Structures, Community Planning, and Amenities. The last heading embraces objectives for 'prohibiting, regulating or controlling the deposit or disposal of waste materials and refuse, the disposal of sewage and the pollution of rivers, lakes, ponds, gullies and the seashore'. Planning authorities were required to draw up the original development plans within 3 years of 1 October 1964, or such longer period as the Minister allowed. Thereafter they are required to review the plan and make such variations thereto as they consider proper, or to make new development plans, from time to time but at least every 5 years. This means that development planning is a relatively new concept in Ireland. The original development plans were generally rather crude and inadequate. Recent plans tend to be more sophisticated and detailed.

Planning authorities are prohibited from including in a development plan any objective the responsibility for the effecting of which would fall on another local authority without consulting the latter.[8] Authorities may make either one development plan for the whole of their area incorporating all of the appropriate mandatory objectives, or two or more development plans, each being for the whole of their area and some one or more of the mandatory objectives, or for part of their area, and all or some one or more of the mandatory objectives.[9] Because of the obvious dangers and inconveniences of allowing eighty-seven different authorities to make eighty-seven different plans, the Act empowers the Minister for the Environment to require the coordination of the plans of two or more planning authorities in certain respects,[10] but the Minister has never exercised this power. The Minister is also empowered to prepare and publish for the use of planning authorities and other persons interested, general instructions in relation to the preparation of development plans and of provisions and clauses usually inserted in such plans, as and when he thinks fit. A number of such circulars have been issued to local authorities.[11] Expert advice and information is available from the technical staff of the Department of the Environment and from An Foras Forbartha.[12]

2.1.2 Procedure for making draft development plans

A draft of a plan, or of a variation to a plan, is prepared by the staff of a planning authority with or without outside assistance. (In some cases, for example, An Foras Forbartha or independent planning consultants are employed.) The proposed draft is submitted to the elected members of the planning authority for their consideration and it may be altered in the light of their reactions to it. Notice of the preparation of the relevant draft must be published in *Iris Oifigiuil* (the Official Gazette) and in at least one newspaper circulating in the area.[13] In certain circumstances, express notice must be served on owners or occupiers of

structures or land where their interests are peculiarly affected.[14] In all cases copies of notices of the preparation of the draft plan or variation and of the written statement comprised therein must be served on prescribed authorities[15] who may have a particular interest in it or who may be in a position to give the council special advice.[16]

Notices published and served must state:

(i)   that a copy of the draft may be inspected at a stated place and at stated times during a stated period of not less than 3 months;

(ii)   that objections or representations with respect to the draft made to the planning authority within the said period will be taken into consideration before the making of the plan or variations;

(iii)   that any ratepayer making objection with respect to the draft may include in his objection a request to be afforded an opportunity to state his case before a person or persons appointed by the planning authority.[17]

The public is thus given a right to inspect the draft plan or variations and to make representations with respect thereto. Plans are not, in practice, presented in such a way as to allow a choice between alternative proposals, nor indeed has there, as yet, been any systematic effort to encourage meaningful public participation in the development planning process.[18] There is a general tendency to see the right of public participation as being more relevant to those whose proprietary interests may be affected by a proposed plan than to the public at large, which has yet to be made aware of the role which it could play in development planning.

A planning authority which has prepared the draft, given the appropriate notices and considered objections and representations made, may decide to amend the draft and make the plan accordingly. If the proposed amendment would constitute a material alteration of the draft displayed, the planning authority must again give public notice of intention to make the proposed amendment, display the draft amendment for not less than *one* month and allow the public to make objections and representations which must be taken into account before the amendment is made.[19] Amendments which are not of a material nature may be made without displaying the amended draft.

When all the prescribed procedures have been properly complied with, the plan or variation or amendment is then adopted and 'made' by the elected members of the planning authority. The making, varying or amending of development plans is a 'reserved function', and must therefore be performed by the elected members of the local authority.[20] Although in law the elected council makes the plan, the officials employed by the council to advise it have the greater influence on its content. Notice of the making of the plan or variation or amendment, as appropriate, must be published in *Iris Oifigiuil* and in at least one local newspaper. This notice must state that a copy of the relevant plan as made, varied or amended is available for public inspection at a stated place and at stated times.[21] Copies are kept available in the offices of planning authorities during office hours. Any member of the public is entitled to obtain a printed copy of a development plan or of an extract therefrom on payment of a fee not exceeding the reasonable cost of making it.[22]

In *Finn* v. *Bray Urban District Council*[23] it was held that the prescribed procedure for making development plans was a mandatory one. In that case, failure to publish the prescribed notice in respect of a draft plan as amended resulted in a declaration that the adoption of the development plan was of no effect.

The Minister for the Environment has power to require the variation of a development plan in respect of matters and in a manner specified by him, and a planning authority is obliged to comply with such a request.[24] In so far as can be ascertained, the Minister has never formally requested a variation of a plan.

2.1.3 Significance of development plans

In essence, development plans serve as land-use allocation maps providing a basis for development control. Planning authorities are statutorily obliged to 'take such steps as may be necessary for securing the objectives which are contained in the provisions of the development plan'.[25] Local authorities, although not required to obtain planning permission for development in their own areas,[26] are prohibited from effecting any development which materially contravenes their development plan.[27] Development plans operate as a framework within which planning permissions are made and granted in that planning authorities must have regard to the provisions in their plans when considering planning applications; development which materially contravenes a provision in a development plan will not normally be permitted. In the exceptional circumstances when such development is to be permitted special procedures must be observed,[28] except where permission is granted on appeal by An Bord Pleanala.

The obligation to make development plans compels local authorities to focus on the proper planning and development of their areas; this is a prime justification for making development plans. Whether this objective is achieved or not is a different matter. It has, for example, proved extremely difficult to keep plans up to date and responsive to demands for change. The usefulness of plans as guides to future decision-making has been reduced by the lack of forward and long-term planning by many authorities. In recent years, also, it has become increasingly obvious that there is a grave need for the articulation of regional as distinct from local planning procedures—especially where questions of industrial location, the siting of national utilities and pollution control are concerned. At present there is little cooperation between planning authorities for rational development, particularly in urban areas. Successive Ministers for the Environment have not exercised their powers to require the necessary coordination of plans.

Although in theory plans may contain objectives for water pollution control and for the control of waste disposal,[29] in practice these objectives are not usually or not adequately stated.

2.1.4 Public participation in development planning

In theory, the underlying policy behind requirements for the notification, display and hearing of views on proposed development plans/variations/amendments is to enable interested members of the public and certain cultural and environmental protection organisations who may be interested in environmental matters and/or who may be affected by proposals in drafts, to participate in the development planning process. In practice most planning authorities have shown little or no enthusiasm for encouraging public participation. Professional planners tend to resent the interference of, and delays caused by 'amateurs'; most politicians prefer to keep their knowledge of the planning process to themselves and to distribute it, when possible, in the guise of political patronage. In addition, by the time public participation is formally invited the planners and the elected representatives have usually developed a commitment to the proposed plan/variation/amendment that inevitably produces a degree of bias against those who suggest alterations. What is in effect sought is a public response to proposals rather than a public input to them and there is little realisation of the nature of the contribution which could be made by ordinary people from their practical and intimate knowledge of local communities and local problems. Alternatives are rarely offered and the public is invited (not expected) to examine and judge a proposal without access to the background information, studies and surveys which influenced the planners and the politicians in making their proposals. This invidious and widespread practice of concealing information from those with a right to participate in the formulation of plans and at whose expense the information has been collected was recently successfully challenged in the US courts who have declared that a prerequisite to the ability to make meaningful comment is to know the basis upon which the rule is proposed.[30] It is to be hoped that the Irish courts will soon be invited to make a similar declaration.

Planning authorities appear to be quite assiduous in complying with the letter as distinct from the spirit of public participation requirements. Of thirty-five authorities (seventeen urban district councils and eighteen borough and county councils) who responded to a questionnaire on public participation in development planning circulated by this writer in 1978, only one—Dublin County Council—indicated that it had displayed its draft for more than the prescribed minimum of 3 months. All county boroughs and county councils displayed their drafts in more than one place but only one urban district council did this. Of the thirty-five respondents, Dublin County Council made the most strenuous efforts to involve the public in making its development plan. These efforts included displaying the plan in seventeen locations, mounting audiovisual displays and ensuring that qualified personnel, and even county councillors, were available to explain the plan to interested members of the public. Of the other authorities, only nine took measures in addition to those prescribed in the Acts to involve the public. The most frequently taken non-mandated measures were:

(i)   ensuring extra press publicity;

(ii)   explaining the draft to residents and other community associations.

Newspapers do not generally manifest a great deal of interest in development planning; only two planning authorities (Dublin and Cork County Councils) stated that national newspapers had taken 'some' interest in their draft plans. Local newspapers tended to display more interest—four authorities considered that they had taken a 'great' interest in the draft displayed, thirteen 'some', four 'little' and one authority considered that its local newspaper had evinced no interest at all. It is difficult to discover the extent of general public interest in development planning because of the lack of statistical records kept of numbers who inspected drafts or who made representations or objections with respect thereto. Only thirteen of the thirty-five respondents kept such records. One borough council reported that not one person had inspected its draft plan. The draft displayed by Dublin County Council was, as might be expected, inspected by the greatest number (21,141); drafts displayed by other county councils attracted an average of about 50 people while the numbers who inspected drafts displayed by urban district councils varied from 1 to 250. There appears to be greater interest in acquiring copies of the development plans when made than in making them (for

example, Cork County Council sold about 1000 copies of its development plan). The greatest number of plans sold by an urban district council was 70. Typical purchasers were those engaged in the construction industry and residents' associations.

Attitudes taken by planning authorities towards public participation in development planning were curious. All thirty-five respondents professed to be in favour of increased public participation, but four qualified their replies by stating that they needed more financial resources and thirteen said that they would welcome participation 'if it were more constructive', thereby implying that in their experience public participation was not constructive. Eight authorities considered that participation in their development planning process was good; nine considered that it was fair and fourteen said that it was bad. There appeared to be a widely-felt view (eighteen out of thirty-five respondents) that the prime motivation for participation was concern about the impact of proposals on proprietary interests.

Public participation appears to have very little impact on the content of plans eventually made. Eighteen authorities reported that their draft had not been altered at all as a result of public comments while fifteen replied that their drafts had been changed in a few respects. The draft plans of urban district councils appear to have met with an extraordinary degree of public acceptance—thirteen out of sixteen replies stated that their drafts had not been changed in any material respect in response to public opinion. Of the three urban district councils who changed their drafts, two made zoning changes and the third altered a proposed traffic route. Twelve county and borough councils changed their drafts. In four cases, proposals relating to roads were varied; in seven, zonings were changed; in six, proposals relating to amenities were changed; in three cases, additional buildings were listed for preservation and in two instances buildings listed for preservation were removed from the list. It was generally agreed that the subjects attracting greatest public interest were roads, zoning and amenities in that order. Of the objections made to the provisions in the last draft development plan displayed by Dublin County Council, 87% related to roads, 8% to zoning and 3% to proposals relating to the conservation and preservation of amenities.

Prescribed authorities have a right to receive specific notice of the making of a draft plan together with a copy of the written statement comprised therein.[31] Of the prescribed authorities, An Taisce, the National Monuments Advisory Council, Bord Failte and the Arts Council might be considered as representative of environmental interests. Of the thirty-five respondents, thirty reported that one or more of the prescribed authorities had actively participated in making the plans eventually adopted. The extent to which the various bodies participated varied:

twelve planning authorities considered that of the four prescribed authorities mentioned above, An Taisce had been the most helpful; five considered that the contributions of the National Monuments Advisory Council had been the most helpful and two gave this accolade to Bord Failte. The Arts Council appears to have abandoned its responsibilities in relation to development planning.

Any assessment of the techniques and experiences of public participation in development planning must necessarily be tentative and subjective because of the relative novelty of the concept and the lack of empirical research in the field. The above findings must, however, on any view, be described as depressing. It is submitted that the main reasons for this are inadequate and defective procedures and unenlightened official attitudes to public participation. The difficulties experienced with respect to industrial location and the siting of public developments in recent years may prompt a re-examination of existing techniques and practices.[32]

2.2 SPECIAL AMENITY AREA ORDERS

Under section 42 of the 1963 Planning Act as amended,[33] planning authorities are empowered to make amenity orders where it appears to them that by reason of:

(i) its outstanding natural beauty,

(ii) its special recreational value, or

(iii) a need for nature conservation

an area should be declared an area of special amenity. These orders may state the objective of the planning authority in relation to the preservation or enhancement of the character or special features of the area including objectives for the prevention or limitation of development in the area. The Minister for the Environment may also, if he considers it necessary, require a planning authority to make an amenity order in relation to a specified area.

Where the functional areas of two planning authorities are contiguous, either authority may, with the consent of the other, make an amenity order in respect of an area in, or partly in, the functional area of the other. Such orders may be revoked or varied by subsequent order and there is an obligation to review any order from time to time and at least once in every period of 5 years for the purpose of deciding whether a revocation or amendment of the order is desirable.

As soon as may be after making an amenity order, the relevant planning authority must publish in one or more newspapers circulating in the area to which the order relates a notice: .

(i)   stating the fact of the order having been made, and describing the area to which it relates;

(ii)   naming a place where a copy of the order and of any map referred to therein may be seen during office hours;

(iii)   specifying the period (not being less than 1 month) within and the manner in which objections to the order may be made to the planning authority;

(iv)   specifying that the order requires confirmation by the Minister and that, where any objections are duly made to the order and are not withdrawn, a public local inquiry will be held and the objections will be considered before the order is confirmed.

When the period for making objections has expired, the planning authority must submit the order for the Minister's confirmation with objections which have been duly made and not withdrawn. If no objection has been made, or if objections made are withdrawn, the Minister has a discretion to refuse to confirm the order or to confirm it with or without modifications. If there have been objections which have not been withdrawn, the Minister must cause a public local enquiry to be held and must consider objections made there and the report of the person who held the inquiry. Only then may the Minister exercise his discretion to refuse to confirm the order or to confirm it with or without modifications. Only three special amenity area orders have been made to date—by Dublin County Council, Dublin Corporation and Dun Laoghaire Borough Council. All three orders are in respect of the Dublin Bay area. Dublin County Council has not yet published its order. There were no unwithdrawn objections to the order made by Dun Laoghaire Borough Council. There were three unwithdrawn objections to the order made by Dublin Corporation. A lengthy public inquiry to consider these was held in May 1978 and the Minister's decision refusing to confirm the order was announced in November 1981. One of the main reasons which inspired the public pressure leading to the making of the three orders was an unsuccessful attempt to locate an oil refinery in the Dublin Bay area. A special amenity area order may be annulled by resolution of either House of the Oireachtas.

In deciding whether to permit a development or the retention of a structure, planning authorities have regard, *inter alia*, to any special amenity area order relating to their areas.[34] Permission will not normally be granted for a development which contravenes an order. A planning authority which intends to permit a material contravention of any provision in a special amenity area order must follow special procedures before a decision permitting such a development is taken.[35] Certain types of development which are normally exempted are not exempted in an area to which a special amenity area order relates. The classes of development involved include mining and certain types of industrial and agricultural developments.[36]

2.3 DEVELOPMENT PROMOTED BY PLANNING AUTHORITIES

Planning authorities are empowered to carry out developments themselves but they may also promote the establishment of developments of benefit to the community at large.[37] Local and planning authorities have extensive powers to purchase land for the purposes of carrying out various statutory duties.[38] Section 77(2) of the 1963 Act states that a planning authority may provide:

(*a*) sites for the establishment or relocation of industries, businesses (including hotels, motels and guest houses), dwellings, offices, shops, schools, churches and other community facilities and of such buildings, premises, dwellings, parks and structures as are referred to in paragraph (*b*) of this subsection,

(*b*) factory buildings, office premises, shop premises, dwellings, amusement parks and structures for the purpose of entertainment, caravan parks, buildings for the purpose of providing accommodation, meals and refreshments, buildings for providing trade and professional services and advertisement structures, buildings or structures for dogs' or cats' homes,

(*c*) any services which they consider ancillary to anything which is referred to in paragraphs (*a*) and (*b*) of this subsection and which they have provided.[39]

Thus, planning authorities are empowered to, and do in practice, provide advance sites for community services and commercial and industrial activities.

2.4 PLANNING PERMISSIONS

2.4.1 The obligation to obtain planning permission

Part IV of the Local Government (Planning and Development) Act 1963, as amended and extended in 1976, provides the legislative framework within which much development of land is controlled.

Section 24 of the 1963 Act provides that permission shall be required in respect of 'any development of land being neither exempted development nor development commenced before the appointed day'. 'Development' is defined in section 3 of the Act as 'the carrying out of any works on, in or under land or the making of any material change in the use of any structures or other land'. It must be noticed that the concept does not include the carrying out of works *over* land. 'Works' are defined in section 2 to include 'any act or operation of construction, excavation, demolition, extension, alteration, repair or renewal'. An artificial meaning was given to the word 'use' in section 2 of the 1963 Act so that in relation to land it does not include the use of the land by the carrying out of any works thereon. The result is that the definition of development embraces two distinct and separate concepts: the carrying out of works and the making of any material change of use. There is no definition of what constitutes a 'material' change of use but section 3 of the 1963 Act does provide, *inter alia*, that where land becomes used for 'the deposit of bodies or other parts of vehicles, old metal, mining or industrial waste, builders' waste, rubble or debris', the use of land shall be taken as having materially changed. Further guidance on what is deemed to constitute a material change in use may be found in judicial decisions and in the practice of planning authorities. For example, a change in kind of use will always be material (e.g. a change from commercial to industrial use); a change in the degree of an existing use may be material if it is very marked. In *Patterson and Patterson* v. *Murphy and Trading Services Ltd.*,[40] Costello J. held that where the nature of activities carried out in a quarry changed in that (a) a different product was produced, (b) a different method of production was employed, and (c) the scale of operations had intensified, there had been a material change of use for which planning permission was necessary. There may, however, be difficulties in establishing that a change is material in the absence of (a) or (b) or especially (c) above. A question which comes to mind in relation to certain industries (e.g. chemical industries) is whether new planning permission is necessary when the operators change the products they produce and the process by which they produce these products but do

37

not intensify the scale of their operations. It is submitted that planning permission could be necessary if the environmental effects of the changes were more than trivial.

2.4.1.1 EXEMPTED DEVELOPMENT

'Exempted' development is defined in section 4 of the 1963 Act as amended in 1976[41] and in Part III of the Local Government (Planning and Development) Regulations 1977. It includes most development by local authorities in their own areas; certain developments by statutory undertakers; certain specified minor developments; changes of use within specified classes of use; certain developments of a temporary nature; development consisting of the use of any land for the purposes of agriculture or forestry; and some developments used for amenity and recreational purposes. In some of the above instances, development is exempted subject to compliance with conditions specified in the Regulations. Development commenced before 1 October 1964 is also exempt from planning control.[42]

Development by 'State authorities', i.e. a Member of the Government, the Commissioners of Public Works in Ireland and the Irish Land Commission, is also considered exempt from the provisions of Part IV of the Planning Acts because of the constitutional doctrine that the State is not bound by a statute unless expressly or implicitly referred to therein.[43] This survival of the royal prerogative was given a republican gloss by the Supreme Court in *Cork County Council and Burke* v. *Commissioners of Public Works*[44] and in *Byrne* v. *Ireland*,[45] and is now justified either on the basis of legislative intention and/or because the State enjoys this exemption in a fiduciary capacity for the public benefit. Section 84 of the 1963 Act does, however, provide that before undertaking the construction or extension of a building (not being a building which is to be constructed or extended in connection with afforestation by the State), a State authority must:

(i) consult with the planning authority to such extent as may be determined by the Minister for the Environment;

(ii) consult with the aforesaid Minister if objections raised by the planning authority are not resolved.

The provision of different and generally less stringent regulatory procedures in respect of developments by many public authorities is a feature of Irish environmental law which has frequently been criticised.

Section 78 of the 1963 Act does envisage that regulations may be enacted requiring planning authorities (only) to give public notice of their pro-

posed developments, to invite objections thereto and to obtain ministerial consent for their developments if objections are not withdrawn. These
regulations have never been made. It must, however, be noted that
provisions similar to those envisaged for these regulations exist in many
statutes empowering public authorities to develop land.[46]

Section 5 of the 1963 Act, as amended, provides that if any question
arises as to what, in any particular case, is or is not development or
exempted development, the question shall be referred to and decided by
An Bord Pleanala from whose decision an appeal may be taken to the
High Court. In 1977 and 1978 thirty-six and twenty-four references
respectively were made to An Bord Pleanala.[47] In *Patterson and Patterson* v. *Murphy and Trading Services Ltd.*,[48] it was held that section 5
did not give exclusive jurisdiction to the Board to determine the question
of what is or is not development or exempted development in all cases
where the question arose: the Board's jurisdiction was limited to resolv-
ing the question when it arose between a planning authority and any
other person in the course of procedures established by the 1963 (but
not the 1976) Act.

2.4.2 Application procedure

Development subject to Part IV of the 1963 Act may not be carried out
except under and in accordance with a permission granted.[49] The procedure governing the submission of planning applications is contained
in Part IV of the Local Government (Planning and Development) Regulations 1977.

2.4.2.1 *LOCUS STANDI* OF APPLICANTS

Although the Act and the Regulations do not specifically limit the *locus
standi* required of an applicant for planning permission or approval, it
appears that, since the Supreme Court decision in *Frescati Estates Ltd.*
v. *Walker*[50] an applicant for permission must have a 'particular degree
of standing'. This was not precisely defined although it was inferred that
'frivolous or perverse applications' would be invalid. Henchy J. also
stated that

> an application for development permission, to be valid, must be
> made either by or with the approval of a person who is able to
> assert sufficient legal estate or interest to enable him to carry out
> the proposed development or so much of the proposed development
> as relates to the property in question.

This purposive definition of the required *locus standi* was an attempt to remedy what was basically bad draftsmanship in the Act and Regulations; the problem which arose in the *Frescati* case (that of a person seeking permission for a development she could not carry out) could easily have been foreseen and had indeed been provided against in similar British legislation. More recently in *Alf-A-Bet Promotions Ltd. v. Bundoran U.D.C.*[51] McWilliam J. in the High Court stated that in his opinion the actual identity of the applicant was immaterial except in cases 'so very unusual as those in the Frescati case'. It is submitted that, having regard to the wording of the Acts and Regulations, McWilliam J.'s approach is the better one.

2.4.2.2 PUBLICITY REQUIREMENTS

The Regulations require that, before an application for planning permission is submitted, notice of intention to make such application must be published either:

(i)   in a newspaper circulating in the district in which the relevant land or structure is situated, or

(ii)   by the erection or by the fixing of a notice on the land or structure.[52]

The notice must state prescribed particulars including the name of the applicant and the location, nature and extent of the development for which permission is sought. A copy of this notice must accompany the planning application.[53] The primary purpose of these requirements is to ensure that the public is given adequate notice that permission will be sought for a particular development so as to enable people to make such representations or objections in respect of the proposed development as they may consider proper. Where notices have not been published in accordance with the Regulations, or are inadequate or misleading for the information of the public, the planning authority has a discretion to require the applicant to publish such further notice in such manner and terms as it specifies.[54] These notification requirements have attracted a good deal of judicial attention. In *Monaghan U.D.C. v. Alf-A-Bet Promotions Ltd.*[55] Henchy J. observed:

> Many people would comment that the permitted modes of notification to the public of planning applications are unfairly inadequate and merit amendment. It is difficult to rebut such a complaint when one realises how easily the present system allows even the most assiduous watcher of the property or of the newspapers to miss the notification of the application.

The courts are insistent that notification procedures, inadequate as they are, be strictly observed. In the same case the Supreme Court unanimously held that, since the published notice was misleading in that it did not state the nature and extent of the proposed development, the planning application was nullified and the applicants could not acquire permission under section 26(4) of the 1963 Act. Again in *Kelleghan Dodd & O'Brien* v. *Corby and Dublin Corporation*,[56] the plaintiffs obtained a High Court declaration that a planning permission was not validly granted because the newspaper notice was inadequate to convey the nature and extent of the proposed development.

2.4.2.3 DOCUMENTATION REQUIRED

A planning application must be accompanied by prescribed documents and particulars.[57] The general purpose of the statutory requirements is to ensure that the planning authority and the public inspecting the documents lodged will be fully informed as to the nature of the development proposed for a particular site or location. Provision is made in the regulations for submitting applications for outline planning permissions. These are a type of permission-in-principle and applications for them need not be as detailed or precise as applications for full planning permissions.[58] However, outline permission, when granted, is conditional on the approval by the planning authority of further details. Applications for planning approval must be accompanied by such further particulars and plans as would be required if full planning permission were sought.[59] A planning authority may request supplementary information but may not use this as a device to defer making a decision on the application. One request only may be made, although further requests may be made if reasonably necessary to clarify the information given in response to the first request.[60] Planning authorities may also invite the applicant to submit revised plans when 'disposed to grant a permission or approval subject to modification' and may grant a permission or approval for the development as modified. The invitation to submit revised plans does not extend the prescribed period for dealing with the application but, if time is running out, the written consent of the applicant for an extension may be obtained.[61]

2.4.3 Procedure on receipt of the application

The provisions regulating the procedure of a planning authority on receipt of a planning application are designed to ensure that the public and certain bodies who may have a particular interest in particular kinds of developments, or developments in particular locations, are adequately notified of proposals for developments. On receipt of the application, the planning authority must publish notice of the receipt and date thereof in a weekly list. Copies of the list must be made available to the elected members of the planning authority and displayed in the offices of the planning authority for at least 4 weeks in a position convenient for public inspection during normal office hours. If for some reason the planning authority wishes to ensure further publicity for an application, or applications generally, it may by resolution decide that the notice be displayed in any other place which it considers appropriate, or published in a local newspaper or made available to any body or group or person likely to be interested.[62] In practice, few planning authorities have exercised their discretion to give more than the prescribed publicity to planning applications. Cork Corporation publishes notice of all applications received by it weekly in the Cork Examiner and six planning authorities send copies of the weekly lists to An Taisce. In certain prescribed cases, a planning authority is obliged to send notice of the application and of the date of receipt thereof to one or more of the prescribed bodies; for example, proposals for developments which would obstruct views or prospects of special amenity value or special interest must be specifically notified to An Taisce, Bord Failte and the Arts Council.[63] Planning authorities may, and in some circumstances must, also require that applications be accompanied by an environmental impact study where the cost of the proposed development would exceed £5m or more and where it would be likely to cause pollution.[64] Particulars of the application must be entered into the register and must be made available for public inspection at the offices of the planning authority during office hours together with brief particulars of the proposed development and (a) any plans, drawings, maps and particulars accompanying it; (b) any further particulars, plans, drawings or maps submitted by the applicant in relation to the application; and (c) a copy, if submitted, of any environmental impact study.[65]

It should be noted that there is no obligation to make available to the public the results of investigations carried out or commissioned by the planning authorities themselves into the possible environmental impact of proposed developments.

2.4.4 The decision on the application

Under section 26(1) of the 1963 Act a planning authority may (a) grant permission unconditionally; or (b) grant permission subject to conditions; or (c) refuse permission.

In considering a planning application, the planning authority is restricted to considering the

> proper planning and development of the area of the authority (including the preservation and improvement of the amenities thereof), regard being had to the provisions of the development plan, the provisions of any special amenity area order relating to the said area

and the matters referred to in section 26(2). Section 24 of the 1976 Act also empowers a planning authority to have regard, when it considers it appropriate, to either or both of the following:

(i)   The probable effect which a particular decision on the matter would have on any place which is not within, or on any area which is outside its area.

(ii)   Any other consideration relating to development outside its area.

Of the eleven matters referred to in section 26(2) as amended, only one, introduced by section 39 of the 1976 Act, i.e. the power to impose conditions for requiring the taking of measures to reduce or prevent noise or vibration, expressly refers to pollution. Nevertheless, planning authorities and the Department of the Environment have interpreted the discretion conferred by section 26 as being wide enough to permit the imposition of conditions to control almost all types of pollution and polluting activities. The Report of the Inter-departmental Environment Committee states that:

> The physical planning system operated by local authorities under the Local Government (Planning and Development) Acts 1963 and 1976 provides a statutory pollution control mechanism of a general nature. Conditions for the control of pollution, whether of water, air or soil, may be attached to planning permission for a project. . . . Permission may be refused if the pollution implications are incompatible with the proper planning and development of the area.[66]

Because of this interpretation of the scope of section 26, there is a widespread tendency to use planning conditions to duplicate or reinforce controls exercised/exercisable under other legislation. Cases frequently arise in which the control of certain matters is not adequately provided for in more specific legislation or where the application of such legislation is less convenient or less certain than the use of planning powers. There is no policy in Ireland that matters should not be dealt with by means of conditions if they are subject to control under other statutes or covered by common law obligations. This is understandable where the common law is concerned, but as a matter of general principle, it is quite wrong that any person should be subject to more than one set of legislative provisions—and penalties—designed to achieve the same end. At present, conditions attached to planning permissions and approvals are the principal form of pollution control in Ireland. The conditions may themselves specify emission or quality standards to be achieved and/or may make the permission conditional on the developer obtaining authorisation for his activities from other pollution control authorities, for example the Alkali Inspector or the Minister for Fisheries. Many of the eighty-seven local planning authorities have not got the expertise, personnel or facilities to assess the environmental impact of, or to specify pollution control measures in respect of, proposed developments. In such cases, advice is usually sought from the Institute for Industrial Research and Standards, An Foras Forbartha, the Department of the Environment or specialist firms in the private sector.

The use by planning authorities of their powers to attach pollution control conditions to the grant of permissions has become more extensive over the years as authorities have become more experienced and the art of development control more refined. To a great extent, this practice arose because of the lack of other mechanisms for controlling pollution. But a perusal of the Planning Acts will indicate that they themselves are singularly inappropriate instruments for pollution control. It is in fact doubtful whether they were intended to fulfil this role to the extent to which they do. Circulars sent by the Minister of the Environment to planning authorities concentrate on the physical, visual and construc-tional aspects of development control and little guidance has been given to local planning authorities on quality or emission standards or suitable codes of practice which might be required to protect the different envi-ronmental media or human health or welfare. Section 26(2) of the 1963 Act does not contain any express reference to pollution control. The reference to noise and vibration controlling conditions which (since 1976) appears in the amended version of section 26(2) was inserted because the attachment of conditions to prevent the intrusion of noise into developments near an airport was declared *ultra vires* in *Frank Dunne Ltd.* v. *Dublin County Council.*[67] One of the reasons for this decision was the fact that the requirements of the condition would have been more appropriately dealt with by building regulations which could be made by the Minister for the Environment under section 86 of the 1963 Act. An extension of this argument might, for example, now be used to advise against the imposition of conditions to control water pollution—other than that preventable or controllable by the siting or design of a development—since the enactment of the Local Government (Water Pollution) Act 1977.

Although planning authorities enjoy a wide discretion to attach condi-tions under sections 26 and 27, there are genuine doubts about the extent to which this discretion may be used. If conditions attached are reasonably designed to ensure respect for objectives properly contained in a development plan or special amenity area order then they acquire a *prima facie* validity being imposed with respect to material considerations. But many development plans do not contain objectives for overall pollution control or in respect of many types of pollution. No development plan, for example, contains objectives for air or water quality standards so that the attachment of conditions prescribing emission or effluent standards for air and water pollution control may be a somewhat questionable exercise. The attachment of conditions to protect sea water from pollution before 1976 was, it is submitted, almost certainly *ultra vires* as the sea is outside the jurisdiction of local authorities and before 1976 planning authorities were expressly 'restricted to considering the proper planning and development of the area of the authority'. This may be one of the reasons which led to the enactment of section 24 of the 1976 Act.

Few planning authorities are sufficiently aware of the legal limits on the *vires* of conditions which they impose and most regard their discretion to attach pollution control conditions to permissions or approvals granted as virtually unlimited. This may prove to be very unwise.

Section 26(4) of the 1963 Act provides that, where an application is made to a planning authority 'in accordance with the permission regulations' for planning permission or approval, and any requirements relating to the application or made under the regulations have been complied with, then, if the planning authority does not 'give notice' to the applicant within the 'appropriate' period, a decision by the planning authority to *grant* the permission shall be regarded as having been given on the last day of that period. The 'appropriate' period is usually 2 months from the date the planning authority receives the application, but it may be 2 months from a later date in certain specified circumstances or if the applicant consents in writing.[68] This particular provision has engaged the attention of the courts on several occasions. In *The State (Murphy)* v. *Dublin County Council*[69] it was held that if notice of a decision was posted to, but did not reach an applicant within the appropriate 2-month period, he could not obtain permission in default. But in *Freeney* v. *Bray U.D.C.*[69a] the High Court held that notice of the decision must be *received* within the appropriate period, otherwise the applicant may be entitled to permission in default. In *Monaghan Urban District Council* v. *Alf-A-Bet Promotions Ltd.*[70] the Supreme Court rejected the applicants' contention that they had obtained permission by default because they had not been notified of the decision within the appropriate period on the grounds that the application submitted was not 'in accordance with the Regulations' as required under section 26(4)(*a*). But in *The State (N.C.E. Ltd.)* v. *Dublin County Council*[71] the High Court made absolute a conditional order of mandamus requiring the planning authority to make a formal grant of planning approval because the council did not give notice of their decision within a 2-month period from the date the application was received. The Council's contention that a letter seeking further information caused the 2-month period to run from a later date failed on the grounds that the letter did not amount to a valid notice for further information or evidence but was more an effort to negotiate changes in the proposed development. This decision was, however, reversed in the Supreme Court.

While the application is being considered by the planning authority, any person has a right to make written objections or representations with respect to it. This facility is, strictly speaking, an informal one but its permissibility is implicitly recognised in article 32(2) of the regulations and it has recently been declared by the Supreme Court in *The State* (Stanford and others) v. *Corporation of Dun Laoghaire*:[71a] it is widely used in practice. If the planning authority is considering permitting a development which would contravene materially the development plan or any special amenity area order, it must do so by resolution supported by a majority which must exceed one third of its members. (Normally, the granting or refusal of a planning permission is an executive function with which the elected members of the planning authority are not concerned.) Other special procedures must also be observed.[72] Planning authorities have been advised that, in deciding whether any development would contravene materially the development plan, 'regard must be had to whether there would be a departure from a fundamental requirement of the plan or whether the development (alone or in conjunction with others) would seriously prejudice an objective of the plan'.[73] Reasons must be given for decisions to refuse permissions or approvals or to attach conditions thereto.[74]

Procedures very similar to those described herein apply in respect of applications for permissions under section 27 to retain structures. These permissions are frequently sought by developers who have carried out a development without permission and who wish to regularise their position.

2.4.5 Appeals

2.4.5.1 APPEALS TO THE PLANNING APPEALS BOARD

Any person may appeal to the Planning Appeals Board against a decision on a planning application. In this situation, *locus standi*[75] to appeal has not been limited in any respect. An applicant/appellant must appeal within 1 month from the date he receives the decision; others must appeal within 21 days beginning on the day the decision was given.[76] The Board is required to determine any appeal as if the application had been made to it in the first instance.[77] It has somewhat wider powers than a planning authority in that it may grant a permission or approval without any extra formality if the proposed development contravenes materially the development plan or any special amenity area order. A deposit of £10 (refundable except where the appeal is vexatious) must be lodged by every appellant.[78] An appellant may request an oral hearing of the appeal, which request the Board may grant or refuse at its discretion. An appellant whose request for an oral hearing has been refused may ask the Minister, within 14 days of notice of refusal of his request, to direct the Board to hold one. The Board is obliged to hold an oral hearing if the Minister for the Environment so directs.[79] The Board has powers to deal with vexatious or unreasonably delayed appeals or references.[80] It may take into account matters other than those raised by the parties to the appeal, provided the parties are put on notice and given an opportunity to comment, and it may make directions as to the expenses of the appeal.[81] The Local Government (Planning and Development) Regulations make detailed provisions as to appeals and the procedures for oral hearings.[82]

All appeals must be made in writing and must contain prescribed particulars. Copies of these documents must be given to each party to the appeal. The planning authority whose decision is appealed against must submit to the Board all the documentation and information obtained from the applicant together with such other documentation or information as the Board may require (much of the documentation will have been available for public inspection in the offices of the planning authority). A party to the appeal (other than a planning authority) must give the Board any document or information in his possession or procurement which the Board considers necessary for the purpose of determining the appeal. If he fails to do this, the Board may deal with the appeal without the document or information. Parties to the appeal may make written observations to the Board on the matter within a time limit specified by the Board. Copies of all such observations must be given to the other parties. The Board may require any party to give public notice in relation to the appeal including the publication of notice of the appeal in a local newspaper. It may also require revised plans or other drawings or proposals modifying the development to which the appeal relates. The Board has the same powers as a planning authority in relation to an application and may, for instance, carry out inspections of proposed sites, etc.

Seven days' notice of the time and place of oral hearings must be given to all parties unless they agree to a shorter period. The hearing may be conducted by the Board or by a person (the 'inspector') appointed for the purpose. The Board or the inspector has a discretion as to the conduct of the hearing, and evidence may be taken on oath, witnesses may be required to attend and to give information requested, and non-parties to the appeal may be heard if the person conducting the appeal consents. If the oral hearing is conducted by an inspector, he must make a written report to the Board, which must include his recommendation(s) relating to the matter at issue. Provision is made for the adjournment or re-opening of oral hearings. The Board is obliged to consider the inspector's report before coming to a decision. Inspectors' reports, oddly, are not made available to the parties to the appeal, but they will be made available if the High Court so directs. If new matters relevant to the appeal or reference come to light after the closing of the oral hearing, but before the Board has taken its decision, a new oral hearing may be held. Great care is taken to ensure that the rules of natural justice are observed and any decision taken in violation of them may be struck down. In *Geraghty* v. *The Minister for Local Government*,[83] the Minister's decision on a planning appeal was set aside because facts which had not been mentioned at the oral hearing and which the plaintiff had no opportunity to controvert had been considered by the Minister in reaching his decision (planning appeals were at that time decided by the Minister).

The Board's decision must comprise a statement specifying the reasons for its decision and must be entered in the register. In 1977 and 1978, there were 3528 and 3551 appeals respectively against section 26 decisions. Of these, 486 and 702 respectively were made by third parties.

2.4.5.2 APPEALS TO THE COURTS

Normally, decisions taken by planning authorities or An Bord Pleanala are final except in so far as the relevant governing legislation provides for an appeal. But a decision of any administrative tribunal including the aforementioned may always be challenged in the High Court on the grounds that it is *ultra vires*, unconstitutional or where there has been an error of law on the face of the record.

The *locus standi* necessary for challenging such a decision has not been precisely defined. An applicant for permission or a party to any appeal would undoubtedly be entitled to sue. Whether suit by any other third party would be entertained is a more difficult matter. In *Law* v. *The Minister for Local Government and Traditional Homes Ltd.*,[84] the plaintiff, who was not an applicant for nor a 'party' to the appeal within the limited meaning given to that expression in the regulations, obtained an order setting aside the Minister's decision on a planning application. He had, however, participated in the proceedings from the outset; his objection has weighed heavily with the inspector who conducted the oral hearing and he lived near the site of the proposed development. He was thus affected in a peculiar and personal manner by the Minister's decision. It is submitted, however, that in planning matters the public interest is so great and the philosophy on public participation in planning legislation such that a suit by any individual other than one suing for frivolous or vexatious reasons, ought to be entertained.

Section 82 of the 1963 Act as amended by section 42 of the 1976 Act provides that proceedings challenging the validity of a planning application, or the Board's decision on an appeal or reference, must be instituted within 2 months from the date the decision was given. This ungenerous time limit may not always be fatal to the institution of proceedings outside the 2-month period. In very exceptional cases patently requiring judicial intervention it may be possible to argue that the section does not apply to void decisions since these are not 'decisions' in the true sense of the word.[85] It may also be possible to bypass the 2-month rule by seeking a remedy under section 27 of the 1976 Act which may have the same practical result as a declaration that a permission is invalid, but which is not, strictly speaking, a 'challenge' of the 'validity' of a decision on a planning application.[86]

2.4.6 Limit of duration of planning permission

Section 29 of the 1976 Act provides for the withering of certain planning permissions 5 years from the date of the coming into operation of the section and, in the case of permissions granted after that date, 5 years from the date of the granting of the permission. Section 2 of the 1982 Act provides for an extension of the 5 year period in certain circumstances.

2.4.7 Revoking or modifying planning permissions

Section 30 of the 1963 Act, as amended,[87] provides that a planning authority may revoke or modify a permission where there has been a change of circumstances relating to the proper planning and development of the area since the previous decision. When a permission is revoked or modified, the change of circumstances which warranted this action must be specified in the decision. There is an appeal to the Planning Appeals Board. The power to revoke or modify may be exercised where the permission relates to the carrying out of works, at any time before those works have been commenced or, in the case of works which have been commenced and which, consequent on the making of a variation in the development plan, will contravene such plan, at any time before those works have been completed. Where the permission relates to a change of use, the power to revoke or modify may be exercised at any time before the change has taken place. Works carried out remain unaffected by a decision to revoke or modify. Compensation may be payable. The decision to revoke or modify must be taken by the elected members of the planning authority.

2.4.8 Enforcement

2.4.8.1 ENFORCEMENT POWERS

Planning authorities have wide powers to enforce the provisions of the Planning Acts. The principal enforcement measures relevant to the subject matter of this book are as follows.

(i) Powers under section 24 of the 1963 Act to prosecute persons who carry out development other than under and in accordance with a permission granted under Part IV of the Act. The maximum penalty on summary conviction is £20 plus £10 for each day on which the contravention continues.

(ii) Powers under section 31 of the 1963 Act, as amended,[88] to enforce planning controls. Under this section a planning authority may serve an enforcement notice requiring that specified steps be taken within a specified period for restoring the land to its condition before the unauthorised development took place or requiring that conditions attached to a permission be complied with. Section 31 notices may require the removal or alteration of any structures, the discontinuance of any use of land or the carrying out on land of any works. If the required steps are not taken as directed, the planning authority may take them at the ultimate expense of the owner of the land. A section 31 notice must be served within 5 years of the 'appropriate period' as defined in the section.

(iii) Powers under section 32 of the 1963 Act, as amended,[89] to serve an enforcement notice requiring specified steps to be taken within a specified time when any condition subject to which a permission for the retention of a structure was granted has not been complied with. If the required steps are not taken as directed, the planning authority may take them and recover the costs from the owner of the structure. This notice must also be served within 5 years of the 'appropriate period' as defined in the section.

(iv)   Powers under section 34 of the 1963 Act, as amended,[90] to prosecute in respect of failure to comply with the requirements of an enforcement notice served under the three preceding sections. The maximum penalty on summary conviction is £250 but a convicted person who does not, as soon as practicable, do everything in his power to ensure compliance with the enforcement notice, shall be guilty of a further offence and liable on summary conviction to a fine not exceeding £50 for each day following his first conviction on which any of the requirements of the enforcement notice (other than the discontinuance of any use of land) remain unfulfilled.

(v)   Powers under section 35 of the Act, as amended,[91] to serve an enforcement notice when authorised development is not being carried out in accordance with permission, requiring that specified steps be taken within a specified time for ensuring that the development is carried out in conformity with the permission. Such notice may require the removal or alteration of any structures, the discontinuance of any use of land or the carrying out on land of any works. If the required steps are not taken as directed, the planning authority may take them and recover the costs from the person on whom notice was served.

(vi)   Powers under section 36 of the 1963 Act, as amended,[92] to serve a notice requiring the removal or alteration of a structure when a planning authority decides that it should be removed or altered. In this instance an appeal against the notice lies to An Bord Pleanala.

(vii) Powers under section 37 of the 1963 Act, as amended,[93] to serve a notice requiring that a use of land be discontinued or that conditions be imposed on the continuance of a use. An appeal against the notice lies to An Bord Pleanala. This power is extremely important from the point of view of pollution control. Compensation is usually payable if it is used except when conditions are imposed 'to avoid or reduce serious air or water pollution or the danger of such pollution'. In the latter circumstance, compensation may be payable if the Minister makes an order under section 58 of the 1963 Act declaring that he is satisfied that it would not be just and reasonable in the particular circumstances that payment of compensation should be prevented.

(viii) Powers under section 25 of the 1976 Act to secure the proper completion of developments by requiring the owner of the land to provide, level, plant or otherwise adapt or maintain open spaces as indicated in the planning application or required by conditions attached to permissions granted. If the owner fails to comply with a written notice, section 25 provides a procedure whereby the planning authority may acquire the land.

(ix) Powers under section 26 of the 1976 Act to prevent unauthorised development at an early stage by serving a warning notice to prevent or stop any unauthorised development (including use) of land or to protect trees or other features which the terms of the permission for development require to be preserved.

(x) In addition to the above powers, planning authorities may also apply to the High Court under section 27 of the 1976 Act, which has in practice proved to be a most effective method of enforcing planning controls. Section 27 provides:

'(1) Where—

(a) development of land, being development for which a permission is required under Part IV of the 1963 Act, is being carried out without such a permission, or

(b) an unauthorised use is being made of land, the High Court may, on the application of a planning authority or any other person, whether or not the person has an interest in the land, by order prohibit the continuance of the development or unauthorised use.

(2) Where any development authorised by a permission granted under Part IV of the 1963 Act has been commenced but has not been, or is not being, carried out in conformity with the permission because of non-compliance with the requirements of a condition attached to the permission or for any other reason, the High Court may, on the application of a planning authority or any other person, whether or not that person has an interest in the land, by order require any person specified in the order to do or not to do, or to cease to do, as the case may be, anything which the court considers necessary to ensure that the development is carried out in conformity with the permission and specifies in that order.'

To date, section 27 has been successfully invoked, *inter alia*, to restrain an auctioneer from using an amplifier in such a way as to cause a nuisance by noise;[94] to prohibit the use of land for the storage or sale of diesel or petroleum products;[95] to prevent the use of premises for holding Dutch auctions[96] and to ensure compliance with planning conditions relating to the disposal of asbestos waste[97] and the treatment of ammonia fumes.[98]

An interesting feature of section 27 is the fact that no time limit is specified within which enforcement action must be commenced in respect of a breach of planning control. Accordingly, the immunity from enforcement action under section 31 of the 1963 Act enjoyed in respect of breaches of planning control which had continued for more than 5 years is now of more limited practical value. In *Galway County Council* v. *Connaught Proteins Ltd.*[99] the High Court granted an order under section 27 prohibiting the use of lands and premises as an animal by-product processing plant even though the land and premises had been so used for more than 5 years. In two other cases, *Dublin County Council* v. *Matra* and *Dublin Corporation* v. *Mulligan*,[100] the High Court expressly re-affirmed that nothing in section 27 nor in any other section of the 1976 Act restricted the time during which a planning authority or other interested person might apply to the High Court under section 27. It was further declared in the latter case that the fact that enforcement proceedings were not brought under section 31 of the 1963 Act did not make unlawful development lawful. But the fact that an applicant for a section 27 order delayed unduly in seeking the order would be taken into account in deciding whether and in what form a section 27 order would be granted. So too would acquiescence in breach of planning control.

The Minister for the Environment may direct planning authorities to serve enforcement notices under sections 31 and 35 of the 1963 Act.[101] Section 28 of the 1976 Act empowers planning authorities to withdraw notices served under sections 30, 31, 32, 35, 36 and 37 of the 1963 Act. This power is intended to facilitate the resolution of difficulties by negotiation. Summary proceedings for breach of section 24 of the 1963 Act or section 26 of the 1976 Act must be commenced:

(i)   at any time within 6 months from the date of the commission of the offence; or

(ii) at any time within 3 months from the date on which sufficient evidence comes to the knowledge of the prosecutor, whichever is the later, but there is an overall limit of 5 years after the commission of the offence.[102]

Penalties for failure to comply with the terms of notices under sections 31, 32, 35, 36 and 37 of the 1963 Act and sections 25 and 26 of the 1976 Act are fines not exceeding £250 plus £100 for each day the offence continues. Contravention of section 26 of the 1976 Act is also punishable by 6 months' imprisonment.[103] Prosecutions may be taken by planning authorities only.

2.4.8.2 EXTENT OF ENFORCEMENT

There are no national statistics on the extent to which Planning Acts are enforced by planning authorities. The following table throws some light on the practice of Dublin Corporation.

Measures taken	1978	1979	1980
Prosecutions for non-compliance with section 31 of 1963 Act notices	48	25	23
Prosecutions for non-compliance with section 26 of 1976 Act notices	2	4	0
Enforcement notices complied with	57	36	36
Warning notices complied with	11	14	38
Other successful actions	84	107	93

In addition, the Corporation brought five successful actions under section 27 of the 1976 Act in 1979 and fourteen in 1980. Despite the statement by the Inter-departmental Environment Committee that the Planning Acts 'provide a statutory pollution control mechanism of a general nature',[104] until recently there have been very few attempts by planning authorities to enforce pollution control conditions in the courts. This is not to state that extra-judicial methods of enforcement have not been employed. The general tendency, however, is to confine enforcement action to physical planning as distinct from pollution offences. Part of the explanation for this could be a local authority practice of relying almost entirely on polluters to monitor their own emissions and discharges.

2.4.9 Individual rights

In no sphere of Irish statute law does the individual enjoy such extensive rights as under the Planning Acts. Most of these rights have been described in some detail in the preceding pages. They include rights:

(i)   to participate in the development planning process;

(ii)   to be notified that a developer proposes to submit a planning application;

(iii) to make objections and representations with respect to development proposals;

(iv)   to appeal against decisions of planning authorities on planning applications to An Bord Pleanala;

(v)   to become parties to the appeal;

(vi)   to request an oral hearing of the appeal;

(vii) to make objections and representations with respect to development proposals to An Bord Pleanala;

(viii) to challenge the validity of planning decisions in the courts;

(ix) to inspect the planning register and documents lodged with planning applications at reasonable times.

It must also be noted that any individual, regardless of his personal or proprietary interests, may enjoy these rights. He does not have to be directly or immediately affected or aggrieved by proposals for developments. In addition, the individual in Ireland has unprecedented rights to enforce planning controls under section 27 of the 1976 Act—whether or not he has an interest in the land.

Section 27 actions have developed into very efficient and speedy mechanisms for enforcing planning controls and also, in cases where a nuisance is caused by an unauthorised use of land, as relatively cheap alternatives to common law nuisance actions. There are already at least two decisions[105] where individuals have availed themselves of section 27 to control noise and dust pollution caused by quarrying and several cases where the threat of instituting proceedings has ensured compliance with the planning code. Once the citizen establishes that:

(i) development is being carried out without permission, or

(ii) an unauthorised use is being made of land, or

(iii) development has not been, or is not being, carried out in conformity with planning permission because of non-compliance with the requirements of a planning condition or for any other reason,

he may invoke the jurisdiction of the High Court to enforce planning controls. Section 27(3) provides that an application for a section 27 order must be by motion.

The discretion conferred on the High Court gives it a great deal of flexibility in deciding whether, and to what extent, it should enforce planning controls at the instigation of an individual.

The most significant factor inhibiting the enormous possibilities of this section as a major instrument for the enforcement of pollution control is the lack of information available as of right to an individual.

2.5 PLANNING BY AGREEMENT

Section 38 of the 1963 Act, as amended by section 39 of the 1976 Act, empowers planning authorities 'to enter into an agreement with any person interested in land in their area for the purpose of restricting or regulating the development or use of land, either permanently or during such period as may be specified in the agreement'. No research has been published indicating to what extent, if at all, this section is used as a pollution control device. It is understood, however, that it has been so used. A 1976 amendment[106] provides that a planning authority in entering into a section 38 agreement may join with a prescribed body. In theory, therefore, the section permits planning authorities to secure the agreement of developers to conditions controlling their use or development of the land, including the generation of pollution, which conditions might not necessarily be properly or legally attachable to a planning permission or approval.[107] The possibility of joining one of the prescribed bodies like An Taisce as a party to the agreement enhances the potential effectiveness of the section and ensures a minimum amount of public participation—albeit indirect—in planning by agreement. Section 38 provides that the agreement may be enforced by the planning authority or any body joined with them but, curiously, no mention is made of enforceability against the planning authority or prescribed body. Presumably the agreement can only be enforced against these bodies if it is supported by consideration.[108] Another potentiality of a section 38 agreement is as a device to ensure that the polluter himself pays the economic costs of pollution control.

2.6 CONTROL OF BUILDING CONSTRUCTION

Section 86 of the 1963 Act empowers the Minister for the Environment to make building regulations the purpose of which would, very broadly, ensure that any new or altered structures are of stable, sound and generally suitable construction and materials; that the space surrounding buildings and building ventilation are of prescribed minimum standards; and that private sewers and drains, sanitary appliances, refuse disposal systems and water suppliers comply with specified standards. Draft building regulations were published in 1976 but have never been brought into force mainly because of vigorous and sustained opposition from the powerful construction industry. There are, therefore, no general legislative controls over building construction in Ireland. Some local authorities (Dublin, Cork, Limerick and Dun Laoghaire) have made building by-laws under various Acts (e.g. Public Health (Ireland) Act 1878, Public Health Acts Amendment Act 1890, Towns Improvement (Ireland) Act 1854) but these are of limited effectiveness and grossly out of date.

There are, however, a number of extra-legal controls over building construction. Many people acquiring new houses are eligible under the Housing Act 1966, as amended, for an exemption from stamp duties and/or a grant provided that the floor areas of the houses do not exceed specified dimensions and that they comply with the minimum standards of the Department of the Environment for a certificate of reasonable value (CRV). In order to obtain a CRV, site and house plans, specifications, planning permissions, septic tank plans (where there is no public sewage system), insulation measures, and in some cases layout plans and building by-law approval must be submitted and approved by the Department of the Environment. In addition, under the Housing Authorities (Loan Charges Contributions and Management) Regulations 1967,[109] as amended, the payment of contributions towards the annual loan charges incurred by a housing authority in respect of the provision of houses and building sites is conditional on the observance by the authority of certain conditions including a condition that 'the dwellings and building sites be approved in accordance with proposals approved by the Minister'. The Department of the Environment has also made available model plans for various types of houses suitable for local authority housing. A housing authority is not required to obtain Departmental approval if it uses these plans.

There are no controls over the construction materials and standards of industrial buildings other than standards, if any, required by the Industrial Development Authority as conditions for grant-aids, and health and safety requirements necessary to comply with the Factories Act 1955, and the Fire Brigades Act 1940.

Notes

1. See 1.4.3.
2. See 1.4.1.
3. S.I. No. 65 of 1977.
4. 1963 Act, s. 19(1).
5. *Ibid.*, s. 19(2).
6. *Ibid.*, s. 19(2) and Third Schedule as amended by 1976 Act, s. 43(1).
7. *Ibid.*, s. 19(2).
8. *Ibid.*, s. 19(4).
9. *Ibid.*, s. 19(5).
10. *Ibid.*, s. 22(2).
11. *Ibid.*, s. 23. See for example Circulars P.D. 2 of 9 December 1974; P.D. 7 of 12 August 1965; P.D. 15 of 8 March 1967; P.D. 16 of 6 June 1967; P.D. 140 of 15 January 1971; P.D. 2 of 14 March 1972.
12. See 1.5.2.
13. 1963 Act, s. 21(1)(*b*).
14. *Ibid.*, s. 21(1)(*c*) and (*d*) as amended by 1976 Act, s. 43(*g*).
15. See 1.6 and 2.1.4.
16. 1963 Act, s. 21(1)(*a*) and Planning Regulations 1976, arts. 5 and 6.
17. *Ibid.*, s. 21(2).
18. See 2.1.4.
19. 1963 Act, s. 21A.
20. *Ibid.*, s. 19(7).

21. *Ibid.*, s. 21(5).
22. *Ibid.*, s. 21(6)(c) and Planning Regulations 1976, arts. 7 and 8.
23. *Finn* v. *Bray Urban District Council* [1969] I.R. 169.
24. 1963 Act, s. 22(3).
25. *Ibid.*, s. 22(1).
26. *Ibid.*, s. 4.
27. *Ibid.*, s. 39. A planning authority must vary the plan if it proposes carrying out a development in conflict with it.
28. *Ibid.*, s. 26(3) as amended by 1976 Act, s. 39(d).
29. *Ibid.*, Third Schedule, Part IV, 11.
30. *Portland Cement Association* v. *Ruchelshaus* 417, U.S. 921(1974).
31. 1963 Act, s. 21(1)(a).
32. Scannell, *Citizen Participation in Environmental Decision-making* (1978), EEC Commission.
33. 1976 Act, s. 40.
34. 1963 Act, ss. 26(1), 27(1).
35. *Ibid.*, s. 26(3) as amended by 1976 Act, s. 39(d).
36. Planning Regulations 1977, Third Schedule.
37. 1963 Act, s. 77.
38. See, for example, Local Government (Ireland) Act 1878, s. 10(1); Local Government (No. 2) Act 1960, ss. 10, 11 and *Leinster Importing Company Limited* v. *Dublin County Council*, H.C. 26.1.1977 (unreported).
39. 1963 Act, s. 77(2) as amended by 1976 Act, s. 43(k).
40. *Patterson and Patterson* v. *Murphy and Trading Services Ltd.*, H.C. 4.5.1978 (unreported). See also *Dublin County Council* v. *Sellwood Quarries Ltd.* [1981] I.L.R.M. 23. *Dublin County Council* v. *Tallagh Block* Co. H.C. 4.11.1981 (unreported).
41. 1976 Act, s. 43(1).
42. 1963 Act, s. 24(1).
43. See Kelly, *The Irish Constitution* (1980), pp. 557–60.
44. *Cork County Council* v. *Commissioners of Public Works* [1945] I.R. 561.
45. *Byrne* v. *Ireland* [1972] I.R. 241.
46. See for example Public Health (Ireland) Act 1878, s. 63, Local Government (Roads and Motorways) Act 1976, ss. 5 and 6.
47. See Annual Reports of An Bord Pleanala, 1977, 1978.
48. *Patterson and Patterson* v. *Murphy and Trading Services Ltd.*, H.C. 4.5.1978 (unreported). See also *Stafford & Bates* v. *Roadstone Ltd.*, H.C. 17.1.1980 (unreported).
49. 1963 Act, s. 24.
50. *Frescati Estates* v. *Walker* [1975] I.R. 177.
51. *Alf-A-Bet Promotions Ltd.* v. *Bundoran U.D.C.*, H.C. 9.5.1977 (unreported). See also *Toft* v. *Galway Corporation* S.C. 30.7.1981 (unreported).
52. Planning Regulations 1977, art. 14.
53. *Ibid.*, art. 17.
54. *Ibid.*, art. 23.
55. *Monaghan U.D.C.* v. *Alf-A-Bet Promotions Ltd.*, S.C. 24.3.1980 (unreported).
56. *Keleghan, Dodd & O'Brien* v. *Corby and Dublin Corporation*, H.C. 12.11.1976.
57. Planning Regulations 1977, arts. 17–20.
58. *Ibid.*, art. 19.
59. *Ibid.*, art. 19(2).
60. *Ibid.*, art. 26. See *The State (N.C.E. Ltd.)* v. *Dublin County Council*, H.C. 3.12.1979 (unreported).
61. *Ibid.*, art. 27.
62. *Ibid.*, art. 24.
63. *Ibid.*, art. 25.
64. See section 12.

65. 1963 Act, s. 41, and Planning Regulations 1977, art. 29.
66. *Report on Pollution Control*, p. 4.
67. *Frank Dunne Ltd.* v. *Dublin County Council* [1974] I.R. 45.
68. 1963 Act, s. 26(4) as amended by 1976 Act, s. 39(*f*). See also Housing Act 1969, s.10.
69. *The State (Murphy)* v. *Dublin County Council* [1970] I.R. 253.
69a. H.C. 16.7.1981 (unreported).
70. *Monaghan U.D.C.* v. *Alf-A-Bet Promotions Ltd.*, S.C. 24.3.1980 (unreported).
71. *The State (N.C.E. Ltd.)* v. *Dublin County Council*, 4.12.1979 (unreported). See also *The State (Conlon Construction Ltd.)* v. *Cork County Council*, H.C. 31.7.1975 (unreported), and *Freeney* v. *Bray U.D.C.*, H.C. 16.7.1981 (unreported).
71a. *The State (Stanford and others)* v. *Dun Laoghaire Corporation*, S.C. 20.2.1981 (unreported).
72. 1963 Act, s. 26(3) as amended by 1976 Act, s. 39(*d*).
73. See Circular P.D. 140 of 15 January 1971.
74. 1963 Act, s. 26(8) as amended by 1976 Act, s. 39(*g*).
75. 1963 Act, s. 26(5)(*a*).
76. *Ibid.*, s. 26(5)(*c*).
77. *Ibid.*, s. 26(5)(*c*).
78. 1976 Act, s. 15.
79. *Ibid.*, s. 16.
80. *Ibid.*, s. 18.
81. *Ibid.*, ss. 17 and 19.
82. Planning Regulations 1977, Part V.
83. *Geraghty* v. *The Minister for Local Government* [1976] I.R. 153.
84. *Law* v. *The Minister for Local Government and Traditional Homes Ltd.*, H.C. 27.5.1974 (unreported).
85. See *Anisminic* v. *Foreign Compensation Commission* [1969] A.C. 147; (1978) 41 M.L.R. 383; (1980) 43 M.L.R. 173; *Freeney* v. *Bray U.D.C.*, H.C. 16.7.1981 (unreported); *The State (Pine Valley Developments Ltd)* v. *Dublin County Council*, H.C. 27.5.1981 (unreported).
86. See 2.4.8. See also *Barrett Builders* v. *Dublin County Council* H.C. 2.5.1979 (unreported).
87. 1976 Act, ss. 24, 28, 39.
88. *Ibid.*, ss. 26, 28, 34.
89. *Ibid.*, ss. 24, 28, 34.
90. *Ibid.*, s. 38.
91. *Ibid.*, ss. 24, 28, 38.
92. *Ibid.*, ss. 24, 28.
93. *Ibid.*, ss. 24, 28.
94. *Dublin Corporation* v. *Clyne & Co.*, Irish Times 15.9.1979.
95. *Monaghan County Council* v. *McCagney*, Irish Times, 3.7.1979.
96. *National Federation of Drapers and Allied Traders Ltd.* v. *Allied Irish Wholesale Warehouses and Others*, Irish Times, 29.11.1978.
97. *Cork County Council* v. *Raybestos (Manhattan) Ltd.*, International Environment Reporter, Vol. 3, No. 7, p. 285.
98. *Cork Council* v. *N.E.T.*, Irish Times, 15.7.1980.
99. *Galway County Council* v. *Connaght Proteins Ltd.*, H.C. 28.3.1980 (unreported).
100. *Dublin Corporation* v. *Mulligan*, H.C. 6.5.1980 (unreported).
101. 1963 Act, ss. 31(1) and 35(1).
102. 1976 Act, s. 30.
103.. 1976 Act, s. 38.
104. *Report on Pollution Control*, p. 4.
105. *Patterson and Patterson* v. *Murphy and Trading Services Ltd.*, H.C. 4.5.1978 (unreported), *Stafford and Bates* v. *Roadstone Ltd.*, H.C. 17.1.1980 (unreported).

106. 1976 Act, s. 39.
107. *Beaconsfield D.C.* v. *Gains* (1974), 234 *Estates Gazette* 749; 1976 *Journal of Planning Law* 732, 738.
108. Alder, J., *Development Control* (1979), pp. 128–137, and Clarke, H.W., 'Enforceability of Development Agreements', N.L.J., 3.7.1975.
109. S.I. No. 71 of 1967; S.I. No. 109 of 1969; S.I. No. 42 of 1970.

3
Atmospheric Pollution

Ireland is fortunate in that it enjoys relatively pure air outside of the two major cities and isolated industrial areas. This is largely because of the geographical location of the country and the direction of the prevailing winds. Nevertheless, it would be unwise to be complacent. Recent studies[1] indicate that air pollution levels in Dublin and Cork have been unacceptably high at times and that pollution in Dublin on occasion exceeds WHO thresholds and only marginally complies with the less stringent norms of the recent EEC Directive on limit and guideline values for sulphur dioxide and suspended particles in the atmosphere. This situation has not been helped by current energy policy which encourages householders to burn solid fuel instead of domestic heating oil.

About one-third of all atmospheric discharges occur in Dublin and one-sixth in Cork, but, as yet, no national inventory of emission sources has been compiled by the Department of the Environment.

The Alkali etc. Works Regulation Act 1906 is the only statute enacted specifically for the control of air pollution from industrial sources but in practice the Local Government (Planning and Development) Acts 1963 and 1976, though not designed as such, are currently the main instruments used for the control of industrial air pollution from developments established since the 1963 Act came into force on 1 October 1964. Regulations to control air pollution have also been made under the Road Traffic Act 1961, the Local Government (Sanitary Services) Act 1962 and the European Communities Act 1972. In addition, some control over air pollution is exercisable under the Public Health (Ireland) Act 1878, and by virtue of conditions attached to various authorisations granted under legislation regulating specific processes and developments. Informal controls over air pollution include conditions attached to grant-aids for new industry and improved effluent treatment plant given by the Industrial Development Authority and for farm modernisation and improvement schemes.[2] The main air pollution difficulties are said to be associated with industries established before 1964 which are not subject to the Alkali etc. Works Regulation Act 1906.

Apart from the few emission standards prescribed in the Alkali etc. Works Regulation Act 1906, and the Control of Atmospheric Pollution Regulations 1970, there are no detailed and uniform emission standards for air pollution from stationary sources. To date no air quality objectives or standards have been defined in Ireland but some contemplated within the EEC will, if accepted, become applicable in Ireland. Control authorities are expected to operate on the basis that standards should be reasonably practicable having regard to all the circumstances relevant to any particular pollution source. This pragmatic approach enables the setting of individual standards for polluting emissions from each particular source although, in practice, there may be a high degree of uniformity in respect of the standards set for some substances because local authorities tend to seek advice from the same sources.

3.1 PUBLIC HEALTH (IRELAND) ACT 1878

Public Health legislation provides for the punishment of a number of statutory nuisances which embrace various types of pollution. The power to abate them is given to local authorities where such action appears to be necessary in order to protect the health of the community. Such 'statutory nuisances' need not necessarily interfere with personal comfort but they can only arise from a private source and they cannot (by definition) arise from public works, such as sewage works. Whether or not a statutory nuisance exists is a question of fact and in this respect the duration of the nuisance may be relevant. Among the statutory nuisances listed in section 107 of the Public Health (Ireland) Act 1878 are:

(i)   'any fireplace or furnace with does not as far as practicable consume the smoke arising from the combustible used therein, and which is used for working engines by steam, or in any mill, factory, dyehouse, brewery, bakehouse, or gaswork, or in any manufactory or trade process whatsoever';

(ii)   'any chimney (not being the chimney of a private dwelling-house) sending forth black smoke in such quantity as to be a nuisance'.

It is a defence to charge of (i) above to prove that the

fireplace or furnace is constructed in such a manner as to consume as far as practicable, having regard to the nature of the manufacture or trade, all smoke arising therefrom, and that such fireplace or furnace has been carefully attended to by the person having charge thereof.

Sanitary authorities are bound to inspect their district for nuisances.[3] On the receipt of any information respecting the existence of a statutory nuisance the sanitary authority is obliged, if satisfied of the existence of the nuisance, to serve an abatement notice on the person by whose act, default or sufferance the nuisance arises or continues.[4] In certain limited circumstances the authority itself may abate the nuisance.[5] If the notice is not complied with, summary enforcement proceedings may be brought in the District Court[6] or, if in the opinion of the sanitary authority summary proceedings would afford an immediate remedy, in any superior court.[7]

3.2 ALKALI ETC. WORKS REGULATION ACT 1906

This Act is administered by the Alkali Inspectorate of the Department of the Environment which is charged with the inspection of all 'works' concerned in the production of substances the manufacture of which may involve the release of 'noxious fumes and smoke'. The Act is partly obsolete, and very limited in scope. It has never been updated to cover new processes. It applies mainly to chemical industries.

3.2.1 Works regulated

3.2.1.1 ALKALI WORKS

Alkali works are works for the manufacture of soda or sulphate of potash, or the treatment of copper ores by common salt or other chlorides whereby any sulphate is formed in which muriatic acid (now called hydrochloric acid) is evolved.[8] Every alkali work must be so carried on as to secure the condensation of the muriatic gas evolved to the extent of 95%, and to such extent that in each cubic foot of air, smoke or chimney gases escaping from the works into the atmosphere there is not more than one-fifth of a grain of muriatic acid.[9] (A cubic foot is to be calculated at 60 °F and 30 inches of barometric pressure.[10]) Contravention of the above requirements is punishable by a fine of up to £50 for a first offence and £100 for every subsequent offence.

In addition, under section 2 of the Act, the owner of every alkali works must use the best practicable means to prevent the escape of noxious or offensive gases by the exit flue of any apparatus used in any process carried out in the works, to prevent their discharge into the atmosphere, and to render them harmless and inoffensive if they are discharged. Any muriatic acid gas which does not exceed the limit imposed by the last preceding paragraph is to be ignored for the purposes of section 2. Failure to comply with the above provisions is punishable by a fine of up to £20 for the first offence and £50 for every subsequent offence together with a further sum not exceeding £5 for each day any subsequent offence continues.[11]

3.2.1.2 SULPHURIC ACID WORKS

Sulphuric acid works are works in which the manufacture of sulphuric acid is carried on by the lead chamber process, in which sulphurous acid is converted to sulphuric acid by oxides of nitrogen and the use of a lead chamber, or by any other process involving the use of oxides of nitrogen.[12] Such works must be carried on so as to secure the condensation of the acid gases of sulphur or of sulphur and nitrogen to the extent that the total acidity of those gases in each cubic foot does not exceed the equivalent of four grains of sulphuric anhydride.[13] Failure to comply with the above requirements is an offence punishable by a fine not exceeding £20 for a first offence, £50 for each subsequent offence and a further sum not exceeding £5 for each day every subsequent offence continues.

3.2.1.3 MURIATIC ACID WORKS

Muriatic acid works are defined as muriatic acid works, or works (not being alkali works as defined in the Act) where muriatic acid gas is evolved either during the preparation of liquid muriatic acid or for use in any manufacturing process; tin plate works, i.e. works in which any residue or flux from tin plate works is calcined for the utilisation of such residue or flux, and in which muriatic acid gas is evolved; and salt works (not being works in which salt is produced by refining salt, otherwise than by the dissolution of rock salt at the place of deposit) in which the extraction of salt from brine is carried on, and in which muriatic acid gas is evolved.[14] Muriatic acid works must be carried on so as to secure the condensation of the muriatic acid gas evolved to the extent that in each cubic foot of air, smoke or chimney gases escaping into the atmosphere from such works there is not more than one-fifth of a grain of muriatic acid. Failure to comply with the above requirements is punishable by a fine of up to £20 for the first offence and £50 for each subsequent offence together with a further sum not exceeding £5 for each day every subsequent offence continues.[15]

3.2.1.4 SCHEDULED WORKS

These are the most important works regulated by the Act. Twenty-one works are listed in the First Schedule to the Act. These include sulphuric and muriatic acid works but not alkali works. Works actually regulated in Ireland include sulphuric acid, chemical manure, nitric acid, sulphate of ammonia and tar works. The owner of any scheduled work must use the best practicable means to prevent the escape of noxious or offensive gases by the exit flue of any apparatus used in any process carried on in the works. He must also prevent the discharge, whether directly or indirectly, of such gases into the atmosphere, and render such gases where discharged harmless and inoffensive. To be ignored for the purpose of deciding whether an offence has been committed are (a) any discharges of muriatic acid gas into the atmosphere where the amount discharged does not exceed one-fifth of a grain per cubic foot of air, smoke or gas, and (b) acid gases from works for the concentration and distillation of sulphuric acid discharged into the atmosphere, where the total acidity of such gases does not exceed the equivalent of one and a half grains of sulphuric anydride per cubic foot of air. The owner of any works operated in contravention of the above provisions is punishable by a fine of up to £20 for a first offence and £50 for each subsequent offence together with a further sum not exceeding £5 each day every subsequent offence continues.[16]

3.2.2 Registration requirements

Every scheduled works and alkali works must be registered with the Alkali Inspectorate and a certificate of registration must be in force. Certificates must be renewed annually. It is a condition of the first registration (or re-registration after 12 months) under the Act that the 'best practicable means' be used for preventing the discharge of noxious and offensive gases into the atmosphere and for rendering them harmless and inoffensive. The expression 'best practicable means' has never been judicially defined in Ireland but a similar expression, 'all practicable measures', used in the Factories Act in England was held to include 'having regard to the state of knowledge at the time and especially the knowledge of scientific people'.[17] The Act itself states that the expression refers not only to the provision and efficient maintenance of appliances adequate for preventing the escape of noxious and offensive gases, but also to the manner in which such appliances are used and the proper supervision by the owner of any operation in which such gases are produced.[18] In practice, the Alkali Inspectorate appears to take a very pragmatic and practical approach: they require that works be efficiently maintained, and that appliances be properly used so as to prevent pollution in so far as is practicable. Account is taken of the cost of any desirable remedial measures; the inspectorate being conscious of a need to preserve a balance between the cost of pollution control measures and the extent of the harm or nuisance involved. The obligation to use the best practicable means is, however, a continuing one and it may entail alterations in plant and/or in its mode of operation as new techniques become available. In practice, means are agreed by the Alkali Inspectorate in consultation with industries affected. The Act does not define emission standards (except for a few acid processes); this is left to the Alkali Inspectorate which prescribes 'presumptive limits' which, if not exceeded, will satisfy them that the best practicable means are being used. These limits are decided in the light of advancing technology and most attention is paid to British, German and American practice. On occasion, the Inspectorate will permit a works to exceed the presumptive limits for justifiable reasons; for example, in 1974 one company in an effort to reduce energy costs sought and received permission to exceed the presumptive limit for its plant so as to burn more sulphur and so produce more steam in a sulphuric acid plant.[19] The Act refers to the registration of works in a register containing 'the prescribed particulars' and the making of an application for a certificate of registration 'in a prescribed manner'. No particulars or manner appear to have been 'prescribed' by statutory instrument so that, in theory at any rate, it appears that there can be no valid applications for certificates of registration.[20]

3.2.3 Enforcement

The Act is enforced by the Alkali Inspectorate. This consists of one or two expert engineering graduates in the Department of the Environment. The number of works registered has never exceeded 20 in any one year. An inspector visits every registered works at least once a year. Some of the inspector's powers have already been described. The most important of these is deciding on applications for certificates of registration. The inspector has all necessary powers for carrying out his duties under the Act; he may:

(i)   enter and inspect any work to which the Act applies;

(ii)   examine any process causing the evolution of any noxious or offensive gas, and any apparatus for condensing such gas, or otherwise preventing the discharge thereof into the atmosphere, or for rendering any such gas harmless or inoffensive when discharged;

(iii)   ascertain the quantity of gas discharged into the atmosphere, condensed or otherwise dealt with; and

(iv)   apply any such tests and make any such experiments, and generally make all such inquiries as seem to him to be necessary or proper for the execution of his duties under the Act.[21]

The owner of any work is obliged, on being requested by the Chief Inspector, to furnish him within a reasonable time with a sketch plan, to be kept secret, of those parts of the production in which any process causing the evolution of any noxious or offensive gas or any process for the condensation of such gas or for preventing the discharge thereof into the atmosphere, or for rendering such gas harmless or inoffensive when discharged, is carried on. He is also obliged to furnish to every inspector all necessary facilities for entry, inspection, examination, and testing in pursuance of this Act. Failure to afford such facilities or obstruction of an inspector in the course of his duty is an offence punishable on summary conviction by a fine not exceeding £10.[22]

The Department of the Environment is obliged to inquire into breaches of the Act occasioning a nuisance on receipt of a complaint from any sanitary authority on information given by any of its officers, or any ten inhabitants of their district, that any work to which the Act applies is carried on (either within or without the district) in contravention of the Act or that any alkali waste is deposited or discharged (either within or without the district) in contravention of the Act, *and* that a nuisance is occasioned thereby to the inhabitants of their district. After the inquiry, the Department may direct an inspector to take such proceedings as it thinks fit and just.[23] In fact the Alkali Inspectorate investigates all complaints made. Individuals are specifically empowered to bring actions for damages for statutory nuisance caused by the owners of works to which the Act applies.[24]

There have been no prosecutions for breach of the Act in recent years— a fact which is not at all surprising since the penalties for breach are totally inadequate. The enforcement powers of the alkali inspector are limited: his power is merely to ensure that the best practicable means are being used, and, while he may prosecute for offences under the Act, he has no legal powers to close down works carried on in flagrant breach of the conditions of registration. He can, of course, refuse to issue a certificate of registration but, in so far as can be determined, he has never done this. There is reason to believe that a good number of works registered under the Act have caused or are still causing air pollution problems. For example, the authors of an IDA survey on pollution in Ireland referred to the fact that there were 'three sulphuric acid manufacturing facilities in the country and all have given or are still giving rise to problems at certain times'.[25] Owners or operators of works are required to monitor their own emissions and to make monitoring records available to the Alkali Inspectorate. All apparently genuine complaints are investigated although it appears that the existence of the Act and of rights available under it are not widely known.

The basic problem with many of the works and processes regulated under the Act is that it is not economically practicable to spend large sums modernising them so as to ensure that they do not cause air pollution. Requiring re-equipment could sometimes result in the closure of a plant and a consequent rise in unemployment statistics. This is a measure which none but the most politically insensitive would hazard.

3.3 LOCAL GOVERNMENT (SANITARY SERVICES) ACT 1962

Section 10(1) of the Local Government (Sanitary Services) Act 1962 empowers the Minister for the Environment, after consultation with other specified Ministers,[26] to make regulations for controlling atmospheric pollution 'for the purpose of securing the cleanliness of the atmosphere or the prevention of danger to health or injury to amenity from atmospheric pollutants'. Section 10(2) provides that regulations may be made for all or any of the following matters:

(i)   Controlling sources of pollution of the atmosphere, including the emission of smoke, dust, grit or gas.

(ii)   Regulating the establishment and operation of:
        (a) trades,
        (b) chemical and other works, and
        (c) processes (including the disposal of waste) which are potential sources of atmospheric pollution from smoke, dust, grit or noxious or offensive gases.

(iii)   Specifying maximum concentrations of specified pollutants in the atmosphere.

(iv)   Measuring emissions of pollutants into the atmosphere.

(v)   Investigating and obtaining information on emissions of pollutants into the atmosphere.

(vi)   Testing, measuring and investigation of atmospheric pollution.

(vii) Regulating potential sources of pollution of the atmosphere from radioactive materials.

(viii) Specifying particular controls of atmospheric pollution for particular areas.

(ix)   Licensing of persons engaged in specified works or processes, being works or processes discharging pollutants into the atmosphere, and prohibiting the engagement in such works or processes of persons other than licensed persons.

(x)   Licensing of premises from which pollutants are discharged into the atmosphere, and prohibiting discharges of pollutants into the atmosphere from premises other than licensed premises.

(xi)   Cancellation or suspension of licenses.

(xii) Imposition of charges for the purpose of the regulations or for services performed thereunder.

The section also empowers the making of regulations conferring jurisdiction on the District Court relating 'to the annulment or confirmation of cancellations or suspensions by sanitary authorities of licences and the direction of sanitary authorities to licence persons or premises in specified circumstances'. Regulations made are enforceable by officers of sanitary authorities or the Department of the Environment who are given necessary powers of entry and inspection. Obstruction of officers in the course of their duties is an offence punishable on summary conviction by a fine not exceeding £20. The licensing systems envisaged by section 10 were never established. Instead, the provisions of the Local Government (Planning and Development) Act 1963 were used to achieve some of the objectives of section 10 although these provisions can only be used prospectively and not, like section 10, for the control of pollution from existing installations or developments. The only regulations made under section 10 were the Control of Atmospheric Pollution Regulations 1970, which are very limited in scope.

3.4 CONTROL OF ATMOSPHERIC POLLUTION REGULATIONS 1970[27]

3.4.1 Scope

These regulations set limits on the length of time during which smoke of varying degrees of darkness may be emitted from premises other than private dwelling houses, and prohibit the emission of smoke, dust, grit, gas or fumes from such premises or from any public place in such a quantity or manner as to be a nuisance to persons in any premises in the neighbourhood.[28]

3.4.2 Operating conditions

Limits are set on the length of time during which dark or black smoke (that is, smoke as dark as, or darker than, shade 2 and shade 4 respectively, on the Ringleman chart) may be emitted from premises other than a private dwelling house. These limits are as follows:

Article 3(1): *Emissions from premises*

Emission	Maximum permissible time
Dark smoke	4 continuous minutes
Darker smoke	8 minutes in the aggregate in any period of 8 consecutive hours
Black smoke	2 minutes in the aggregate in any period of 30 minutes.

Article 3(2): *Emissions from chimneys serving two or more furnaces*

Permitted period of dark smoke in any given 8-hour period	
No. of furnaces served by one chimney	Maximum permitted time
2	16 minutes
3	22 minutes
4	27 minutes

It should be noted that a private dwelling house in the regulations does not include the curtilage or garden of a private dwelling house or a building containing several private dwellings and having a furnace the maximum heating capacity of which exceeds 55,000 British Thermal Units per hour (16.1189 kW).[29] It is an offence to exceed these limits.

3.4.3 Defences

A number of defences are provided for those who exceed the prescribed time limits.

(i)   That the best practicable means have been employed to avoid the contravention.[30]

(ii)   That the emission of the smoke was caused by the lighting up of the fire or furnace from being cold.[31]

(iii)   That the contravention was solely due to some failure of a furnace or of apparatus used in connection with a furnace, that the failure could not reasonably have been foreseen or, if foreseen, could not reasonably have been provided against, and that the contravention could not reasonably have been prevented by action taken after the failure occurred.[32]

(iv)   Where a person is charged with emitting dark smoke for a continuous period exceeding 4 minutes, that the smoke complained of was emitted during soot-blowing.[33]

(v)   Where a person is charged with emitting dark smoke for more than 8 minutes in the aggregate in any period of 8 consecutive hours, that the smoke complained of was emitted during soot-blowing and was not emitted for more in the aggregate in any period of 8 consecutive hours than:
        12 minutes where the chimney is served by one furnace;
        23 minutes where the chimney is served by two furnaces;
        32 minutes where the chimney is served by three furnaces;
        39 minutes where the chimney is served by four furnaces.[34]

(vi)   To a charge of emitting smoke, dust, grit, gas or fumes from premises other than a private dwelling house in such quantity as to be a nuisance to persons in any premises in the neighbourhood, to prove that the best practicable means were taken to minimise the emission complained of.[35]

'Practicable' in the context of the regulations means reasonably practicable having regard, amongst other things, to local conditions and circumstances, financial considerations and the current state of technical knowledge, and 'practicable means' includes the provision and maintenance of plant and the proper use thereof.[36]

3.4.4 Enforcement

The regulations are enforceable by the Department of the Environment and sanitary authorities. In practice, they are enforced by sanitary authorities only. Enforcement officers are given the necessary powers of entry and inspection by the parent Act.[37] Obstruction or interference with an officer in the course of his duty is punishable on summary conviction by a fine of up to £20.[38] In cases where a time limit has been exceeded, or smoke, dust, grit, gas or fumes have been emitted from premises (other than a private dwelling house) in such quantity or manner as to be a nuisance, and where the contravention is likely to continue or recur, the sanitary authority may notify the occupier of the premises of the works what it considers necessary to be done to end the contravention or to prevent its recurrence.[39] Prosecutions may be brought by the Minister or the appropriate sanitary authority for contravention of the regulations. The penalty on summary conviction for contravention or wilful obstruction of the execution of a regulation is a fine not exceeding £100 and, if the contravention or obstruction is continued after conviction, a further fine not exceeding £5 for each day the contravention or obstruction is continued.[40] The Minister for the Environment has suggested to sanitary authorities that proceedings should only be instituted 'as a last resort' and has advised them to seek to resolve problems by cooperating with offenders. Authorities have been advised to inform the Minister when they intend to prosecute so that he may consider the question of obtaining specialist advice to ensure that 'when prosecutions are taken, they will be successful in having the cause of pollution rectified'.[41] Sanitary authorities have adopted an informal procedure of serving notices of breach, followed, on occasion, by notices of intention to prosecute. There have never been any prosecutions under the Regulations. Dublin is the area most likely to suffer from air pollution and the statistics in relation to enforcement of the regulations by the sanitary services section of Dublin Corporation are shown in the following table.

Enforcement of Control of Atmospheric Pollution Regulations 1970

Year	Number of breaches of the regulations	Notices served (verbal and written)
1971	186	169
1972	162	158
1973	170	191
1974	92	92
1975	96	77
1976	107	101
1977	79	53
1978	106	93
1979	123	97

3.5 LOCAL GOVERNMENT (PLANNING AND DEVELOPMENT) ACTS 1963 AND 1976

The scope and effect of these Acts have already been described.[42] Briefly, they enable planning authorities to control through their development plans the location of industry likely to cause air pollution and to refuse permission for, or permit subject to controlling conditions, developments which may generate air pollution. The Minister for the Environment has advised planning authorities to 'consider attaching suitable conditions in relevant cases to ensure that the best practicable means are adopted to minimise the emission of dust and grit' and he has also emphasised the necessity for controlling emissions from fuel-burning plants and chimneys. All authorities were furnished with, and were advised to refer to, memoranda published by the British Ministry of Housing and Local Government on chimney heights and grit and dust from boiler and furnace chimneys.[43] If a planning authority decides that a use of land should be discontinued or that conditions should be imposed on the continuance thereof, it may so require.[44] However, in many cases the planning authority may not do this without attracting liability to purchase the land concerned or to compensate the owner.[45] One of the exceptional cases where compensation may not be payable is where conditions are imposed on the continuance of a use in order 'to avoid or reduce serious air pollution or the danger of such pollution'.[46]

3.6 SULPHUR CONTENT OF CERTAIN LIQUID FUELS[47]

The EEC Council Directive on the approximation of laws of the member states relating to the sulphur content of certain liquid fuels has been partially implemented by the European Communities (Sulphur Content of Gas Oil) Regulations 1977,[48] made under section 3 of the European Communities Act 1972. The Directive sets down EEC requirements for the limitation of the sulphur content of gas oils which would have the effect of ensuring a marked reduction in atmospheric pollution caused by sulphur compounds.

The Directive seeks to attain its objective in two stages: the marketing of type A gas oil is prohibited unless its sulphur compound content, expressed as sulphur, does not exceed 0.5% by weight as from 1 October 1976 and does not exceed 0.3% by weight as from 1 October 1980; the marketing of type B gas oil is prohibited unless its sulphur content, expressed as sulphur, does not exceed 0.8% by weight as from 1 October 1976 and 0.5% by weight as from 1 October 1980. Member states were allowed to define low atmospheric pollution regions and regions where the contribution made to atmospheric pollution by gas oils was low: in these areas the use of type B gas oil (the more polluting type) is permissible.

Ireland has been granted a 5 year exemption before having to pass on to the second stage of the programme for reducing the sulphur content of gas oil, which means that the limits set for 1 October 1980 in the Directive will not be mandatory until 1985. The 5 year exemption was sought because of the large investment which would have been required to enable Whitegate, the only oil refinery in the country, to produce gas oil with a sulphur content conforming with the Directive. Whitegate refined about 55% of the total gas oil consumption in 1978.[49] The implementing regulations provide that the entire State is to be determined as a zone for the use of type B gas oil.[50] The Directive envisages that type B oil shall be used in zones 'where ground level concentrations of atmospheric sulphur dioxide pollution are sufficiently low, or where gas oil accounts for an insignificant proportion of atmospheric sulphur dioxide pollution'. Dublin city could hardly be said to fulfil either of these criteria.

The regulations do not apply to gas oil used in power stations or by shipping or contained in the fuel tanks of motor vehicles entering the State.[51] There is one slight and insignificant difference between the Directive and the Regulations: the exemption in the Directive in respect of gas oil contained in the fuel tanks of inland waterway vessels has not been granted in Ireland.

The regulations are enforceable by the Department of Energy. The penalty for breach on summary convictions is a fine not exceeding £500 and/or six months' imprisonment. Any person who obstructs or interferes with an authorised person seeking entry to any place where gas is kept for the purpose of examining it and taking samples of gas oil is liable on summary conviction to a fine not exceeding £250.[52]

3.7 LEAD IN PETROL

The EEC Council Directive on the approximation of the laws of the member States concerning the lead content of petrol[53] provides that petrol with a lead content not exceeding 0.64 grams per litre may be placed in the Irish market for a period of 5 years from 1 January 1981 whereas in other member states the maximum permitted lead content of petrol placed on the market from 1 January 1981 will be 0.40 grams per litre. The derogation sought by Ireland has been justified by the Minister for the Environment on the grounds that

the large investment which would have been required to enable the Whitegate refinery to produce petrol with a lead content conforming with the directive would have placed the continued operation of the refinery in jeopardy. In addition, imports of crude oil would have to be increased to produce petrol with a lower lead content and there would also be an increase in petrol consumption, estimated at 10 per cent.[54]

The Minister has, however, stated that

> as a result of the monitoring of levels of atmospheric lead pollution at three stations in the Dublin area, the derogation which Ireland obtained from the terms of the Directive will not result in levels of atmospheric lead above the 2 microgram per cubic metre level proposed in the draft EEC Directive on air quality standards for lead.[55]

Studies carried out by the Institute for Industrial Research and Standards in the autumn of 1981 indicate, however, that the Minister has not been correctly appraised of the true position with respect to lead levels in Dublin.

3.8 MOTOR VEHICLES

3.8.1 Implementation of EEC Directives

Although the initial aim of the European Commission in the harmonisation of requirements as to the technical characteristics of motor vehicles was to remove technical barriers to trade, this aim was expanded to ensure that damage to the environment from the use of mechanically propelled vehicles is minimised. The EEC has adopted a number of Directives which require, *inter alia*, that motor vehicles be so constructed as to minimise emissions of pollutants. Since motor vehicles are not manufactured—as distinct from assembled—in Ireland, the implementation of these Directives was a relatively straightforward process. The European Communities (Motor Vehicle Type Approval) Regulations 1978,[56] made under section 3 of the European Communities Act 1972, implement the provisions of certain EEC Directives which relate to the type approval of motor vehicles, trailers and components. The relevant Directives are 70/156/EEC of 6 February 1970 as amended by 73/350/EEC of 7 November 1973 and by 77/212/EEC of 8 March 1977 on sound level and exhaust systems; 70/220/EEC of 20 March 1970 as amended by 74/290/EEC of 28 May 1974, by 77/102/EEC of 30 November 1976, and by 78/665/EEC of 14 July 1978 on air pollution by gases from vehicles with positive ignition engines; and 72/306/EEC of 2 August 1972 on atmospheric pollution by diesel engined motor vehicles.

Moreover, the European Communities (Vehicle Type Approval) Regulations 1980[57] provide that components of mechanically propelled vehicles and trailers which comply with the provisions of the scheduled EEC Directives and continue in use so to comply, shall not be regarded as infringing existing statutory requirements. Under the European Communities (Motor Vehicle Type Approval) Regulations 1978,[58] the Institute for Industrial Research and Standards is empowered to issue type approvals to manufacturers of products used in vehicle construction provided the products conform to EEC requirements. The regulations also provide for the conditions subject to which type approval may be issued; for the circumstances in which type approval may be cancelled or suspended; for the manner in which type approved products may be identified; and for the right of aggrieved manufacturers to appeal to the Minister for Industry and Commerce against any decision on a type approval. Contravention of the regulations is punishable on summary conviction by a fine not exceeding £300 and/or up to 6 months' imprisonment. The regulations are enforceable by the Department of Industry and Commerce.

3.8.2 National requirements

The Road Traffic Act 1968 empowers the Minister for the Environment to make regulations to control, *inter alia*, the importation, supply and fitting of substandard parts and equipment for vehicles;[59] to introduce a type approval scheme;[60] and to give effect to the European agreement concerning Uniform Conditions of Approval for Motor Vehicle Equipment and Parts, 1958.[61] The Minister has never made any of these regulations but some of the objectives of the 1968 Act have been achieved by the European Communities (Motor Vehicle Type Approval) Regulations 1978.

3.8.2.1 CONTROLS OVER DESIGN AND CONSTRUCTION

Controls over the design and construction of motor vehicles marketed in Ireland are contained in the European Communities (Motor Vehicle Type Approval) Regulations 1978. Other controls are contained in Part II of the Road Traffic Acts 1961 and 1968, and in the Road Traffic (Construction, Equipment and the Use of Vehicles) Regulations 1963, as amended.[62] The 1961 Act places responsibility for compliance with the relevant regulations solely on the user (or owner) of the vehicle, but this responsibility (and controls) were extended by the 1968 Act to ensure that vehicles and vehicle parts were supplied in proper condition.[63] The Road Traffic (Construction, Equipment and Use of Vehicles) Regulations 1963 apply in relation to the use of vehicles in public places and, in so far as they attempt to deal with air pollution, they provide that 'every vehicle shall be so constructed as to prevent, to such extent as is reasonably possible, the emission of smoke, visible vapour, noxious gases or offensive odours'; that 'every vehicle using solid fuel be fitted with an efficient appliance for the purpose of preventing the emission of sparks or grit and also with a tray or shield to prevent ashes or cinders from falling to the ground'; and that a vehicle 'propelled by a compression ignition engine shall not be fitted with a device designed to facilitate the starting of the engine by causing it to be supplied with excess fuel unless:

(a) the device is so designed that it cannot cause the engine to be supplied with excess fuel after the engine has been started, or

(b) the device cannot be readily operated by a person while he is carried in the vehicle'.[64]

3.8.2.2 CONTROLS OVER MAINTENANCE

The Road Traffic (Construction, Equipment and Use of Vehicles) Regulations 1963, as amended, provide in respect of the maintenance of vehicles that:

(i)   Every vehicle and trailer, and all parts and equipment of every vehicle and trailer, shall be maintained in good and efficient working order, and shall be such and so maintained that no danger is liable to be caused thereby.

(ii)   Every vehicle and trailer shall be maintained in such a condition that there will not be emitted any smoke, visible vapour, grit, sparks, ashes, cinders or oily substances, the emission of which could be prevented by the taking of any reasonable steps or the exercise of reasonable care, or the emission of which might cause damage to person or property or endanger the safety or health of any other user of the public place in consequence of any harmful content contained therein.

(iii) Where a vehicle with a compression ignition engine is fitted with a device designed to facilitate starting by causing the engine to be supplied by excess fuel, the device shall be maintained in such a condition that it does not cause the engine to be supplied with excess fuel while the vehicle is in motion.[65]

Vehicles, other than public service vehicles, are not required to be tested at intervals in order to ensure that they are roadworthy.

3.8.2.3 CONTROLS OVER USE

The following are offences under the Road Traffic (Construction, Equipment and Use of Vehicles) Regulations 1963, as amended.

(i)   To use a vehicle in a public place so that there is emitted therefrom any smoke, visible vapour, grit, sparks, ashes, cinders or oily substances, the emission of which could be prevented by the exercise of reasonable care.

(ii)   To use, cause or permit the use of a device designed to facilitate starting so as to cause it to supply the engine with excess fuel while the vehicle is in motion.[66]

3.8.2.4 ENFORCEMENT

Enforcement of the Road Traffic Acts 1961 and 1968, and the regulations made thereunder, is the responsibility of the Garda Siochana. There are no statistics on the extent to which the regulations referred to in this work are enforced. Enforcement depends on visual observations and subjective judgement by the Gardai. The Report of the Inter-departmental Environmental Committee on Pollution Control recommended 'the substitution of an objective system of monitoring and control in relation to polluting emissions from vehicles for the subjective system which is at present the basis of the regulations.'[67] The penalty for breach of the regulations is a fine not exceeding £20 for a first offence and £50 for a second or subsequent offence.[68]

3.9 AIRCRAFT AND SHIPS

There are no national legislative controls over air pollution caused by aircraft, although all aircraft landing in Ireland must comply with International standards.

For the purposes of the Control of Atmospheric Pollution Regulations 1970, a ship or vessel lying in any river, harbour, or other water not within the district of a sanitary authority is deemed to be within the district of the nearest adjoining sanitary authority. Prosecutions under these Regulations may be brought in respect of prohibited emissions from any ship or vessel.[69]

3.10 MISCELLANEOUS CONTROLS

A number of statutes enacted to regulate specific developments introduce authorisation requirements under which there is a discretion to attach air pollution control conditions to authorisations granted. These include the Petroleum and Other Minerals Development Acts 1940–1979,[70] the Smelting Act 1968, and the Nuclear Energy Act 1971. In addition, some extra-legal control may also be imposed by conditions attached to grant-aids awarded by the Industrial Development Authority[71] and by the Department of Agriculture in the administration of the Farm Modernisation Scheme.[72]

3.11 INDIVIDUAL RIGHTS

Apart from his rights at common law the individual has a number of rights under some of the statutes referred to in this chapter. Under the Public Health (Ireland) Act 1878, any person aggrieved by a statutory nuisance (which may include air pollution), or any two inhabitant householders of a district may complain of the existence of the nuisance to the sanitary authority which is *obliged*, if satisfied of the existence of the nuisance, to serve a notice requiring the abatement thereof and, if this does not have the desired effect, to prosecute the person deemed responsible for the nuisance.[73] A person aggrieved by a statutory nuisance or the inhabitant of a sanitary district or the owner of any premises in the district may also prosecute for a statutory nuisance.[74]

Under the Alkali etc. Works Regulation Act 1906, any ten inhabitants of a district may complain to their sanitary authority that any work to which the Act applies is being carried on in contravention of the Act and that a nuisance is occasioned thereby to any of the inhabitants of their district. Where the sanitary authority in turn passes on this complaint to the Department of the Environment, the Department is obliged to 'make such inquiry into the matters complained of, and after the inquiry may direct such proceedings to be taken by the inspector as they think fit and just'.[75] The Alkali Act also provides for actions by individuals in cases of contributory nuisances, i.e. cases where a nuisance (public or private) arises from a discharge of any noxious or offensive gas or gases caused by several discharges from several sources.[76] Any person injured by such nuisance may proceed against any one or more of the owners and may recover damages from each defendant-owner to the extent of the contribution of that defendant to the nuisance, notwithstanding that the act or default of that defendant would not separately have caused a nuisance.

The rights of the individual under the Local Government (Planning and Development) Acts 1963–1976 have already been described in detail.[77] Of particular importance is his right under section 27 of the 1976 Act to ensure that new development causing air pollution is undertaken and carried out in accordance with permission granted under the Acts. The provisions of other statutes referred to in this section are enforceable by designated authorities only.

3.12 MONITORING[78]

Regular monitoring of air quality for sulphur dioxide and smoke is carried out by local authorities in Dublin, Cork, Limerick, Waterford and Galway—the major urban areas. Dublin Corporation also monitors lead in the atmosphere at three stations. The Meterorological Service of the Department of Tourism and Transport operates a programme of monitoring the chemical content of precipitation at nine stations as part of the International Meteorological Institute network. The Meteorological Service also monitors radioactivity in the atmosphere at two stations. The Electricity Supply Board operates twenty-four stations in the Dublin area which monitor sulphur dioxide and smoke and also monitor pollution levels in the vicinity of its generating stations. Monitoring networks in Dublin adhere to British Standard 1747 (1963) procedure for sampling the ambient air quality.

The EEC Decision of 24 June 1975[79] establishing a common procedure for the exchange of information between the surveillance and monitoring networks, based on data relating to atmospheric pollution caused by certain compounds and suspended particulates, has been implemented by the Department of the Environment.

The monitoring of industrial sources of air pollution is carried out, if at all, by dischargers and, on occasion, by local authorities. The Department of the Environment, in a circular,[80] has instructed local authorities that

> in dealing with new industries which may give rise to an air pollution problem, local authorities should ensure that adequate arrangements are made for the monitoring of air emissions from the projects and for assessing the findings.

For this purpose, authorities were advised that in some cases it might be necessary to attach conditions to planning permissions 'that specified monitoring be carried on by the developers and information supplied on an on-going basis to the local authority'. Alternatively, or in addition, authorities were advised to undertake monitoring themselves in the vicinity of particular projects. The circular also contained a list of industries considered as potential sources of significant air pollution. Local authorities were advised to make arrangements for in-plant and/or outside monitoring to ensure that adequate information is obtained on air emissions and their effects where there was a *prima facie* reason to believe that industries were causing significant air pollution.

There is very little published information of the outcome of monitoring programmes. In 1971–74, the Institute for Industrial Research and Standards investigated 70 cases of actual or alleged air pollution in the vicinity of single sources.[81] The results of these investigations showed that of the 77 cases examined

> 7 were confirmed as air pollution cases causing injury to man or his environment; 51 were confirmed as being a 'source of nuisance', though no injury was being caused and 19 were found to be causing neither injury nor nuisance.

The IIRS believe that their investigations covered the bulk of all actual air pollution incidents in Ireland in the period 1971–74. From this the IIRS concluded 'that only seven instances of injury or damage have occurred in recent years is a firm indication of how relatively unpolluted is Ireland's air'.[82] Such a conclusion is not at all surprising when the criterion for dissatisfaction with air pollution levels adopted was that of actual injury to man and his environment and when investigations were confined to sources already identified as sources of 'actual or alleged' air pollution. The IIRS also found that about 40% of the incidents of air pollution nuisance stem from food and food by-product processing. The manufacture of various building and construction materials accounts for a further 25% of the incidents, while the chemical industry is responsible for about 10%.[83]

Notes

1. Bailey, M.L., 'A Study of Dublin Air Quality 1970–75' (1978), National Board for Science and Technology, Dublin. Walsh, J. J. and Bailey, M. L., 'Air Pollution— Impacts and Control' (1978), National Board for Science and Technology, Dublin, pp. 18–29.
2. See 1.3, 1.5 and 4.8.
3. Public Health (Ireland) Act 1878, s. 109.
4. *Ibid.*, s. 110.
5. *Ibid.*, s. 110.
6. *Ibid.*, ss. 111, 121.
7. *Ibid.*, s. 123.

8. Alkali etc. Works Regulation Act 1906, s. 27(1).
9. *Ibid.*, s. 1.
10. *Ibid.*, s. 16.
11. *Ibid.*, s. 2.
12. *Ibid.*, First Schedule para. 1.
13. *Ibid.*, s. 6(1).
14. *Ibid.*, First Schedule para. 8.
15. *Ibid.*, s. 6(2).
16. *Ibid.*, s. 7.
17. See *Adsell* v. *K.&L. Steelfounders and Engineers* [1953] 2 All E.R. 320.
18. Alkali etc. Works Regulation Act 1906, s. 27(1).
19. See Annual Report of the Alkali Inspector, 1974.
20. Alkali etc. Works Regulations Act 1906, s. 9.
21. *Ibid.*, s. 12(1).
22. *Ibid.*, s. 12(2), (3), (4).
23. *Ibid.*, s. 22.
24. *Ibid.*, s. 23, see below.
25. *A Survey of Pollution in Ireland* (1976), Industrial Development Authority, p. 10.
26. Local Government (Sanitary Sources) Act 1962, s. 10(3).
27. S.I. No. 156 of 1970.
28. *Ibid.*, arts. 3, 5.
29. *Ibid.*, art. 2.
30. *Ibid.*, art. 3(3)(*a*).
31. *Ibid.*, art. 2.
32. *Ibid.*, art. 3(3)(*b*).
33. *Ibid.*, art. 3(3)(*c*).
34. *Ibid.*, art. 3(4).
35. *Ibid.*, art. 3(5).
36. *Ibid.*, art. 4(2).
37. See 3.3.
38. Local Government (Sanitary Services) Act 1962, s. 10(5).
39. Control of Atmosphere Pollution Regulations 1970, art. 6.
40. Local Government (Sanitary Services) Act 1962, s. 10(6).
41. Circular L 4/71 of 2 February 1971.
42. See Chapter 2.
43. Circular ENV 7/73 of 11 June 1973.
44. Local Government (Planning and Development) Act 1963, s. 37.
45. *Ibid.*, ss. 37(8), 29, 61.
46. *Ibid.*, s. 61 as amended by Local Government (Planning and Development) Act 1976, s. 41(*f*).
47. OJ L 307, 27 November 1975.
48. S.I. No. 361 of 1977.
49. *Dail Reports*, Vol. 320, Col. 641, 1 May 1980.
50. The European Communities (Sulphur Content of Gas Oil) Regulations 1977, art. 4.
51. *Ibid.*, art. 6.
52. *Ibid.*, art. 5.
53.. OJ L 197, 22 July 1978.
54. *Dail Reports*, Vol. 320, Col. 641, 1 May 1980.
55. *Ibid.*
56. S.I. No. 305 of 1978.
57. S.I. No. 41 of 1980.
58. S.I. No. 305 of 1978.
59. Road Traffic Act 1968, s. 9.
60. *Ibid.*, s. 8.
61. *Ibid.*, s. 14.

62. S.I. No. 190 of 1963.
63. Road Traffic Act 1968, ss. 8, 9, 14, 17.
64. Road Traffic (Construction, Equipment and Use of Vehicles) Regulations 1963, art. 30.
65. *Ibid.*, art. 34.
66. *Ibid.*, art. 90.
67. *Report on Pollution Control*, p. 26.
68. Road Traffic Act 1961, ss. 11(4), 103; Road Traffic Act 1968, s. 8.
69. See 3.4.
70. See 5.6 and 1.3.
71. See 1.5.1.
72. See 1.3.3 and 4.8.
73. Public Health (Ireland) Act 1878, ss. 109, 110, 111.
74. *Ibid.*, s. 121.
75. See 3.2.3.
76. Alkali etc. Works Regulation Act 1906, s. 23.
77. See 2.4.9.
78. See Coffey, J., *Air Pollution—Impacts and Control* (1978), National Board of Science and Technology, Dublin, pp. 130–132; Bailey, M. L. and Walsh, J. J., 'Monitoring of Air Pollutants in Dublin', *Irish Journal of Environment Science*, Vol. 1 (1979), pp. 26–35.
79. OJ L 194 25 July 1975.
80. Circular ENV. 22/74 of 10 September 1974.
81. *A Survey of Pollution in Ireland* (1976), Industrial Development Authority, p. 10.
82. *Ibid.*
83. *Ibid.*

4

Pollution of Inland Waters

In general the quality of Irish inland waters is believed to be good. An Foras Forbartha reports that of the 121 major rivers it surveyed in the last decade, 75% of the 2900 kilometres surveyed continue to have satisfactory water quality. Moreover, there has been no increase in the numbers of seriously polluted rivers (6%) in the last 10 years although there has been a two-fold increase (10% to 22%) in stretches of moderately polluted rivers due mainly to more widespread eutrophication. Of 50 major lakes surveyed, 15 were found to be enriched—8 excessively so and 7 moderately.[2]

The law on water pollution in Ireland has recently been reformed and, to an extent, codified. The Local Government (Water Pollution) Act 1977 is now the main instrument for the control of water pollution. But water pollution controls in other Acts of lesser scope have been continued in force. The more important of these so far as inland waters are concerned include the Public Health (Ireland) Act 1878 and the Local Government (Planning and Development) Acts 1963 and 1976.

Water pollution control is primarily, though not exclusively, the responsibility of local authorities. Some control is also exercised by Fisheries Boards, the Minister for Fisheries and by private individuals. Many new controls have been inspired and will continue to be inspired by the necessity for complying with measures prescribed under EEC law.

Various EEC Council Directives relevant to water quality have been implemented or are in the process of being implemented. These are Directives 76/160/EEC on bathing water, 75/440/EEC concerning the quality required of surface water intended for the abstraction of drinking water, 80/68/EEC on the protection of groundwater against pollution caused by certain dangerous substances, 76/464/EEC on pollution caused by certain dangerous substances discharged into the aquatic environment of the Community, and 78/659/EEC on the quality of fresh waters needing protection or improvement in order to support fish life. Apart from these, there are no other mandatory water quality or treatment standards in Ireland. Instead, water pollution control authorities have a discretion to require compliance with whatever quality or emission standards they think fit when granting permissions of various kinds for discharges to waters. Some degree of uniformity in standards required may be expected of control authorities following the recommendations of the Technical Committee on Effluent and Water Quality Standards published in *Memorandum No. 1 on Water Quality Guidelines.*

4.1 PUBLIC HEALTH (IRELAND) ACT 1878

Section 19 of the Public Health (Ireland) Act 1878 provides that sanitary authorities may not discharge sewage or filthy water into any natural stream or watercourse, or into any canal, pond or lake unless the discharge is freed from 'all excrementitious or foul or noxious matter'. Section 77 prohibits the contamination by gas washings of any stream, reservoir, aqueduct, pond or place for water or any drain or pipe communicating therewith. The penalty for this offence is £200 and a further sum of £20 for every day the offence continues. Proceedings may be brought by the appropriate sanitary authority or by the person whose waters were polluted. Section 78 empowers sanitary authorities with the sanction of the Attorney General to take proceedings in their own name or in the name of any other person with the consent of that person to prevent water pollution by sewage whether the sewage originates inside or outside their district.

Section 107 enacted that a statutory nuisance shall exist where 'any pool, ditch, gutter, watercourse, privy, urinal, cesspool, drain or ashpit is so foul or in such a state as to be a nuisance or injurious to health'. Sanitary authorities are obliged under the Act to inspect their districts for such nuisances,[2] and on receipt of a complaint about a nuisance from a person aggrieved by it or any two inhabitants of the district, or one of their own officers, or a police officer, and if satisfied of the existence of the nuisance, to take steps to ensure that the nuisance is abated.[3] Any person aggrieved by a nuisance, or the inhabitant of a sanitary district, or the owner of any premises in the district may also prosecute for a statutory nuisance.[4]

4.2 FISHERIES (CONSOLIDATION) ACT 1959

Sections 171 and 172, as amended,[5] of the Fisheries (Consolidation) Act 1959 established licensing and certification schemes respectively under which licences and certificates could be granted by the Department of Fisheries for discharges to waters. Section 171(1) provides that:

Any person who:

(a) steeps in any waters any flax or hemp, or
(b) throws, empties, permits or causes to fall into any waters any deleterious matter

shall, unless such act is done under and in accordance with a licence granted by the Minister under this section, be guilty of an offence. . . .

'Waters' for the purposes of sections 171 and 172 means 'any river, lake, watercourse, estuary or any part of the sea'. Deleterious matter is defined as

any substance (including explosive liquid or gas) the entry or discharge of which into any waters is liable to render those or any other waters poisonous or injurious to fish, spawning grounds or the food of any fish or to injure fish in their value as human food, or to impair the usefulness of the bed and soil of any waters as spawning grounds or their capacity to produce the food of fish.

The Minister for Fisheries is empowered, after consultation with the Minister for Industry and Commerce in respect of applications for licences for industries, and the Minister for the Environment in respect of applications for licences by sanitary authorities in relation to sewage schemes, to attach conditions to any licence and, after like consultation, to revoke any licence.

Licences are usually granted for limited periods, are subject to review and stipulate, *inter alia*, and where appropriate, monitoring requirements to be fulfilled and standards to be complied with by licensees. Since the schemes established under the Act are not compulsory, the number of licences held under section 171 has never exceeded 30. At present licences are held by sanitary authorities, food processing and chemical firms. The IDA and planning authorities sometimes require recipients of grant-aids and planning permissions respectively to obtain section 171 licences.[6]

Section 172 of the Act, as amended, provides that it is an offence to discharge 'deleterious liquid' which is contained or conveyed in a receptacle which is within 30 yards of any waters, to any waters. The owner of a receptacle who has provided 'suitable means' for the prevention of deleterious discharges may apply to the Minister for a certificate specifying the suitable means and the manner in which they are to be used. If he observes the terms of the certificate he may not be convicted of an offence under section 172. No certificate of suitable means has ever been issued.

The Local Government (Water Pollution) Act 1977 envisaged the repeal of sections 171 and 172[7] but they have been continued in force because they provide a mechanism for controlling the dumping of wastes at sea and because of administrative delays in implementing the Local Government (Water Pollution) Act 1977. An interesting feature of the sections is the fact that they apply to deleterious discharges to water irrespective of the identity of the discharger and thus accord no special treatment to discharges made by the public sector. The sections are enforceable by the Garda Siochana, Fisheries Boards[8] and the Minister for Fisheries. A licence has never been revoked. The penalty for breach of section 171 is a fine not exceeding £500 and/or 6 months' imprisonment. The penalty for breach of section 172 is the same plus a fine not exceeding £50 for each day (subject to a maximum of £600) the offence continues after conviction. In practice, the sections were enforced almost exclusively by Boards of Fisheries Conservators in District Courts. The following table, adapted from the Dail Reports,[9] indicates the extent to which sections 171 and 172 were enforced by Boards of Conservators.

Year	Total number of prosecutions	Number of prosecutions against local authorities
1974	31	6
1975	25	7
1976	59	9
1977	40	12
1978	55	25

These figures relate to prosecutions taken by 14 of the 17 Boards of Conservators.

The Minister for Fisheries has also made by-laws[10] under section 9 of the 1959 Act and section 33 of the Fisheries (Amendment) Act 1962, prohibiting the discharge of any effluent from sand or gravel washing plants into any waters in the State without his consent. The Minister may attach such conditions as he thinks fit to his consent and may vary any conditions attached subject to giving 1 month's written notice to the person concerned.

4.3 LOCAL GOVERNMENT (PLANNING AND DEVELOPMENT) ACTS 1963 AND 1976

The scope and effect of these Acts have already been described.[11] Briefly, they enable planning authorities to control through their development plans the location of industry likely to cause water pollution, and to refuse permission for, or to permit subject to controlling conditions, developments which may generate water pollution. Planning authorities have, in theory at any rate, a complete discretion as to what, if any, water quality and effluent standards they require.[12]

4.4 LOCAL GOVERNMENT (WATER POLLUTION) ACT 1977

4.4.1 Scope

Under this Act, local authorities have primary, but not exclusive responsibility for ensuring the preservation, protection and improvement of water quality. Because of the fundamental changes which this Act wrought on Irish water pollution law, section 33 thereof provided that different provisions might be brought into force on different dates. Much of the Act came into operation on 1 May 1977[13] but sections 4 and 16, which require the licensing of discharges to waters and sewers, did not come into operation until 1 October 1978 and 1 January 1979 respectively.[14] Commencement orders have now been made in respect of all sections of the Act other than sections 25 and 34.

In section 1 of the Act 'waters' are defined so as to include:

(i) any (or any part of any) river, stream, lake, canal, reservoir, aquifer, pond, watercourse or other inland waters, whether natural or artificial;

(ii) any tidal waters;

(iii) where the context permits, any beach, river bank, and salt marsh or other area which is contigious to anything mentioned in (i) or (ii), and the channel or bed of anything mentioned in (i) which is for the time being dry, but does not include a sewer.

The inclusion of an aquifer in the above definition and the further definition of an aquifer as 'any stratum or combination of strata which stores or transmits sufficient water to serve as a source of water supply' means that the Act applies to the vast bulk of (and perhaps all) groundwaters. This is the first time that legislation has been enacted for the direct protection of groundwaters. Furthermore, the Act applies to both inland and sea waters, the expression 'tidal waters' in section 1 being defined to include 'the sea and any estuary up to high water mark medium tide and any enclosed dock adjoining tidal waters'. Pollution is nowhere defined but 'polluting matter' is very widely defined and it includes

> any poisonous and noxious matter, and any substance (including any explosive, liquid or gas) the entry or discharge of which into any waters is liable to render those or any other waters poisonous or injurious to fish, spawning grounds or the food of any fish, to injure fish in their value as human food, or to impair the usefulness of the bed and soil of any waters as spawning grounds for their capacity to produce the food of fish, or to render such waters harmful or detrimental to public health or to domestic, commercial, industrial, agricultural or recreational uses.[15]

This definition of polluting matter is an amalgam of the definition of deleterious matter in the Fisheries (Consolidation) Act 1959[16] and part of the definition of water pollution in section 3 of the Illinois Environmental Protection Act 1970.

4.4.2 Prohibition of the entry of polluting matter to waters

Section 3 of the Act provides that 'a person shall not cause or permit any polluting matter to enter waters'. It should be noted that the section creates two separate offences: (1) causing and (2) permitting polluting matter to enter water. The difference between the two offences can be seen in *Alphacell Ltd.* v. *Woodward*[17] and *Price* v. *Cormack*.[18] It appears that 'causing' involves some active operation or chain of operations involving as a result pollution; 'permitting' on the other hand involves a failure to prevent. Section 3(1) is concerned only with the entry of polluting matter to waters—not with pollution itself—so that it is not necessary to prove that pollution has actually occurred. Once it is established that the matter which entered the waters is *liable* to damage the legitimate uses of those or any other waters (e.g. waters downstream or outside the mixing zone), then an essential ingredient of the offence is proved. The recognition of defined quality standards for the various protected uses of water would appear to be essential for proving that they are liable to be detrimentally affected by polluting matter.

It is a defence to prove that the person charged with the offence took all reasonable care to prevent the prohibited entry.[19] 'All reasonable care' should be judged in the light of good management practices, available technology and scientific knowledge where, for example, industrial effluents are involved; and in the light of current good agricultural practice where agricultural effluents are involved.[20]

The prohibition in section 3(1) does not apply to trade or sewage effluents controlled under other provisions in the Act; the entry of matter from vessels, from apparatus for transferring any matter to or from vessels, or from marine structures to tidal waters; and discharges and deposits authorised under various specified enactments.[21]

It does, however, apply to trade and sewage effluents exempted by the Minister for the Environment under section 4(10), i.e. small amounts of domestic sewage and trade-effluents discharged by sanitary authorities other than from a sewer, except where these discharges comply with quality standards prescribed by the Minister under section 26(1). The Minister has not yet prescribed such standards.

Prosecutions may be brought by a local authority, a Fisheries Board, the Minister for Fisheries or any other person. The penalty on summary conviction is a fine not exceeding £250 together with, in the case of a continuing contravention, a fine not exceeding £100 for each day the contravention is continued and/or 6 months' imprisonment. The penalty on conviction on indictment is £5000 together with £500 for every day on which the contravention is continued and/or 2 years' imprisonment.[22]

4.4.3 Discharges of trade and sewage effluents to waters

4.4.3.1 THE OBLIGATION TO OBTAIN A LICENCE

Section 4 of the Act provides that (subject to exceptions) a person shall not, after such date as may be fixed by the Minister,

> discharge or cause or permit the discharge of any trade effluent or sewage effluent to any waters except under and in accordance with a licence under this section.

The date fixed by the Minister was 1 October 1978.

A trade effluent means an effluent discharged from premises used for carrying on any trade or industry (including mining) but the expression does not include domestic sewage and storm water; trade includes agriculture, aquaculture, horticulture and any scientific research or experiment. Sewage effluent is treated or untreated sewage. Premises include land, whether or not there are structures on the land. The section does not apply in respect of discharges of trade or sewage effluents unless they enter waters from 'any works, apparatus, plant or drainage pipe' used for their disposal to waters. Section 4 therefore is directed at point sources only. Run-off of effluents, slurry, fertilizers or biocides from land are thus not controlled under section 4—though they are covered by the general prohibition in section 3.[23]

4.4.3.1.1 Exempted discharges

Section 4 does not apply to discharges:

(i)   to tidal waters from vessels or marine structures;

(ii)   from a sewer as defined in the Local Government (Sanitary Services) Acts 1878 to 1964;

(iii)   exempted by regulation made under section 4(10).

Once again, there is evidence here of the general tendency in Irish environmental legislation to provide different regulatory procedures for public and private activities. Effluents from sewers are discharged for the most part by sanitary authorities and are thus exempt from the licensing requirements of section 4. They must, however, comply with any relevant quality or treatment standards for effluents or waters which may be prescribed by the Minister for the Environment under section 26.[24] The Minister has not yet prescribed standards under section 26.

The Local Government (Water Pollution) Regulations 1978[25] exempt two classes of effluents from section 4(1):

(i)   domestic sewage not exceeding 5 cubic metres in volume in any 24 hour period which is discharged to an aquifer from a septic tank or any other disposal unit by means of a percolation area, soaking pit or other method;

(ii)   trade effluent discharged by a sanitary authority in the course of the performance of its powers and duties, other than from a sewer.[26]

However, these exempted discharges are, by virtue of section 3(5)(a), subject to the general prohibition in section 3 unless they comply with any relevant standard prescribed by the Minister for the Environment under section 26.

Licences are granted by local authorities which, in the present context, means county councils and borough corporations but not urban district councils or town commissioners.

4.4.3.1.2 Existing discharges

This Act applies to new and existing discharges but section 5 makes special provision in respect of the latter: once a licence application is made for an existing discharge (as defined in section 5) before 1 October 1978, and any information required by any regulations under section 6 is furnished, the applicant may continue to make discharges without contravening section 4(1) (but not section 3(1)) until such time as the local authority grants or refuses a licence. Since there is no time limit specified within which a local authority is obliged to decide on a licence, this could be indefinitely. Apparently section 5 was motivated by an appreciation of the difficulties which might be experienced by existing dischargers in complying with new requirements, and local authorities use the delay in deciding on applications as a lever to encourage dischargers to make progressive improvements in the quality of effluents discharged.

4.4.3.2 APPLICATION PROCEDURE

The procedures regulating the submission of applications for licences are as contained in Part II of the Local Government (Water Pollution) Regulations 1978. They are, in essence, very similar to the procedures regulating the submission of applications for planning permissions already described in some detail.[27]

4.4.3.2.1 Publicity requirements

An applicant for a licence under section 4 must publish in a local newspaper notice of his intention to apply for the licence and of certain prescribed particulars.[28] The local authority may require the applicant to publish further notice of his intention on the same grounds as a planning authority may in respect of a notice of intention to make a planning application.[29] The obligation to publish notice of intention does not apply where the discharge is an existing discharge or where it is a discharge from a development for which planning permission or planning approval has been granted in the five years preceding 1 October 1978: public notice of the grant of a licence in such cases will be given by the local authority.[30]

4.4.3.2.2 Documentation required

A licence application must be accompanied by the following.

(i)   Such plans and particulars in duplicate and such other particulars as are necessary to describe the premises, drainage system and any works, apparatus or plant from which the effluent is to be discharged and to identify the receiving waters and the point of discharge.

(ii) Particulars of the nature, composition, anticipated temperature, volume and rate of discharge of, and the proposed treatment of, the effluent and the period or periods during which the effluent is to be discharged.

(iii) In the case of a trade effluent, a general description of the process or activity giving rise to the discharge.

In addition, a licence application for an existing discharge must state that the discharge is such.[31] Local authorities have powers to request further information and, if the applicant fails or refuses to comply with such a request, to procure the requested information (at the partial or total expense of the applicant) or to decide on the application without it.[32]

4.4.3.3 PROCEDURE ON RECEIPT OF THE APPLICATION

A local authority is obliged to make the application and accompanying documentation available at its office for public inspection during normal office hours until the application or any appeal relating thereto is determined.[33] There are no provisions (as there are in respect of planning applications) which require the local authority to publish weekly lists of applications received, or to notify prescribed bodies of the receipt of any application. Neither is a local authority expressly empowered to ensure further publicity for the application, although it probably has implied powers to do so. Curiously, no provision is made whereby arrangements can be made to respect the confidentiality of information submitted.

4.4.3.4 THE DECISION ON THE APPLICATION

Section 4(3) provides that:

A local authority may at its discretion refuse to grant a licence under this section or may grant such a licence subject to such conditions as it thinks appropriate and specified in the licence.

In considering whether to grant a licence a local authority must have regard to the objectives contained in any water quality management plan, and a licence may not be granted in respect of the discharge of an effluent which would not comply with, or would result in the waters to which the discharge is made not complying with, standards prescribed by the Minister for the Environment under section 26.[34] As yet, no water quality management plans or standards have been made or prescribed.

However, local authorities have been circulated with Water Quality Guidelines, one of the purposes of which is to provide them with 'guidelines on water and effluent quality for their assistance in dealing with development proposals which may affect water quality'.[35] Local authorities must also ensure compliance with standards in EEC Directives when issuing licences.

Without prejudice to the generality of section 4(3), conditions attached to licences may:

(a) relate to—

    (i) the nature, composition, temperature, volume, rate, method of treatment and location of discharge, the periods during which a discharge may be made or may not be made, the effect of a discharge on receiving waters and the design and construction of outlets for a discharge;

    (ii) the provision and maintenance of meters, gauges, other apparatus, manholes and inspection chambers;

    (iii) the taking and analysis of samples, the keeping of records and furnishing of information to the local authority;

    (iv) the prevention of a discharge in the event of a breakdown of plant;

(b) require defrayment of or contribution towards the cost incurred by the local authority in monitoring the discharge;

(c) specify a date not later than which any conditions shall be complied with; and

(d) require the payment to the local authority which granted the licence of a charge or charges prescribed under, or calculated in accordance with the method prescribed under regulations made by the Minister for the Environment.[36]

There is no time limit within which a local authority must give a decision on a licence but as the procedural aspects of the planning and water pollution control systems are so alike it is envisaged that, at least in the case of applications for licences for new discharges, decisions should be made within 2 months of receipt by the local authority of a properly completed licence application.

Section 24 of the Act empowers the Minister to make regulations requiring local authorities, sanitary authorities and Fisheries Boards to consult with such persons and in such manner in relation to the exercise of such powers and duties under the Act as may be prescribed. To date no consultation procedures have been prescribed, but local authorities have been advised to develop consultative arrangements administratively.[37] In particular, all section 4 applications must be forwarded to the Department of Fisheries. Other bodies to be consulted include County Committees of Agriculture where pollution from agricultural activities is involved; Health Boards on matters relating to public health; the Industrial Development Authority in relation to industrial promotion issues; and tourism and conservation interests 'as necessary'. By the end of 1979, fewer than a dozen licences had been granted. About 300 licences were granted in 1980. Since then the Minister for the Environment has urged local authorities to adopt a definite time scale for dealing with applications in hand. Priority is to be given to dealing with applications in respect of major and highly polluting discharges. Every effort is to be made 'to appreciate the problems of particular industries': in this respect, and when setting time schedules within which prescribed or necessary effluent quality standards should be met, local authorities were advised when considering suitable conditions to take into consideration issues such as 'viability, capital expenditure, increased operating costs, process change, availability of land and the degree of pollution being caused'.[38]

While the application is being considered, any person may submit written objections or representations to the local authority. This facility is, strictly speaking, an informal one but its permissibility is implicitly recognised in article 11(1)(c) of the Regulations.

Once it has come to a decision, the local authority must, if the decision is a positive one, transmit the licence to the applicant, or, if the decision is negative, give notice of refusal to the applicant. Notice of the decision must also be given to any third party who submitted written objections or representations, or, alternatively, notice may be given to third parties by publication of the decision in a local newspaper. The notice must, *inter alia*, inform the applicant and the parties of their right to appeal against the decision under section 8 of the Act. A model form of licence is contained in the Second Schedule to the Regulations.[39]

A register must be kept of all licences granted under section 4.[40] Conditions attached to a licence are binding on any person discharging or causing or permitting the discharge of effluent to which the licence relates.[41] The licence lapses if no discharge authorised by it is made within 3 years or if such a discharge ceases for 3 years.[42]

Discharging a controlled effluent except under and in accordance with a licence is an offence. Prosecutions may be taken by a local authority, a Fisheries Board, the Minister for Fisheries or any other person.[43] It is a good defence to a prosecution for an offence under any Act, other than the Local Government (Water Pollution) Act 1977, that the act constituting the alleged offence is authorised by a section 4 licence.[44] The penalties on conviction are the same as for breach of section 3(1).[45]

4.4.3.5 REVIEW OF LICENCES

Section 7 of the Act provides that licences will be subject to review at intervals of not less than 3 years. A licence may, however, be reviewed at any time with the consent of the person making the discharge or where the local authority has reasonable cause for believing that the discharge is a significant threat to public health or where there has been an unforeseen material change in the condition of the receiving waters. It must also be reviewed if standards are prescribed by the Minister for the Environment under section 26.

A local authority must give notice to the discharger and the public of its intention to review a licence. The notice must, *inter alia*, state that representations relating to the review may be made within 1 month.[46] The discharger may be required to submit plans or other particulars necessary for the purposes of the review and these must be made available for public inspection at the local authority offices until the review or any appeal relating thereto is determined.[47] If plans or particulars requested are not supplied within 3 months, the review may be completed without them.[48] Following the review, the local authority may amend or delete conditions attached to a licence or attach new conditions, and if appropriate, issue a revised licence.[49] Where conditions in a licence which permits the discharge of certain dangerous substances to groundwaters have not been complied with, EEC Council Directive 80/68/EEC requires that permission to discharge these substances be withdrawn. Notice of the decision on the review containing, *inter alia*, information as to the recipient's right to appeal the decision must be given to the discharger and to any person who submitted written representations.[50]

4.4.3.6 APPEALS TO THE PLANNING APPEALS BOARD

Any person may appeal to the Planning Appeals Board against a decision on a licence application under section 4 or in relation to the amendment or deletion of conditions attached to, or to the attachment of new conditions to a licence on review under section 7.[51] Appeals relating to the grant or refusal of a section 4 licence must be made within 1 month from the date of the grant or refusal: appeals relating to the decision on review under section 7 must be made within 1 month of the decision.[52] Part IV of the Local Government (Water Pollution) Regulations 1978 prescribes the procedures to be followed on appeal. In general, they are very similar to those governing planning appeals.[53] One substantial difference, however, is that a person refused an oral hearing by the Board has no right to appeal the decision to the Minister for the Environment. Section 8 of the Act provides that after consideration of an appeal (for which no time limit is specified) the Board may either refuse the appeal or

> give appropriate directions to the local authority concerned relating to the granting or revoking of a licence or the attachment, amendment or deletion of conditions, and, where such directions are given, the authority shall, as soon as may be after receipt of the directions, comply with them.

The Board must notify every party to an appeal of its decision, and every notification to persons other than a local authority must specify the nature of the decision, including any directions given to the local authority relating to the granting or revoking of a licence or the attachment, amendment or deletion of conditions.[54] There is no obligation to give reasons for the decision or for the imposition of conditions (if any). A local authority must notify the holder of a licence when it complies with directions of the Board.[55]

4.4.4 Monitoring by the discharger

Section 4(5) of the Act enumerates amongst the conditions which may be attached to a licence granted under that section, conditions 'requiring the provision and maintenance of meters, gauges, other apparatus, manholes and inspection chambers' and conditions requiring 'the taking and analysis of samples, the keeping of records and furnishing of information to the local authority'. It is likely that such conditions will always be attached to licences if appropriate, as local authorities have not the finances, facilities or manpower to continuously monitor discharges themselves. Section 23 of the Act requires any person who is abstracting water from or discharging effluent or other matter into any waters to furnish information to the local authority concerning the abstraction or discharge within the time specified, which may not be less than 14 days.

4.4.5 Water quality management plans

Section 15 of the Act enables a local authority to make a water quality management plan for any waters situated in its functional area or which adjoin that area. Making a plan is mandatory if the Minister for the Environment so directs. Plans may cover any waters in or adjoining the functional area of the local authority but may include the sea only to the extent that the Minister, after consultation with the Minister for Fisheries, shall approve. Plans must contain such objectives for the prevention and abatement of water pollution and such other provisions as the local authority considers necessary. A plan may not contain any provision which is inconsistent with quality and treatment standards prescribed under section 26.

The procedures for making plans is contained in Part IV of the Regulations. Public notice must be given of intention to make a plan; the plan must be made available for public inspection for not less than 3 months; written representations may be made by any person with respect thereto and copies of the plan, when made, must be made available for public inspection and for sale at a reasonable cost. Plans may be revised or replaced from time to time. Making, revising or replacing a plan is a reserved function and must be performed by the elected local authority. Copies of all current plans must be furnished to the Ministers for the Environment and Fisheries, to adjoining local authorities and to sanitary authorities and Fisheries Boards in the functional area of the local authority. The Minister has power to require a local authority to revise a plan and to require two or more local authorities to make a joint plan or to coordinate their plans.

No water quality management plan has yet been made. It is understood, however, that An Foras Forbartha are preparing a number of plans for local authorities on a river catchment basis.

A local authority in considering whether or not to grant or refuse a licence for discharge to waters, and a sanitary authority in considering whether to license discharge to sewers, must have regard to objectives contained in any relevant water quality management plan.[56]

4.4.6 Registers of licences and water abstractions

Section 9 of the Act and Part V of the Regulations require the keeping of registers of licences for discharges of (1) trade or sewage effluents to waters (section 4), (2) trade effluent or other water to sewers (section 16), and (3) water abstractions other than abstractions which do not exceed 25 cubic metres in any 24 hour period. The register must be in a prescribed form and contain prescribed particulars and it must be made available for public inspection.

4.4.7 Enforcement

Provisions of the Water Pollution Act may be enforced by a number of authorities and, in some instances, by any person whether or not that person has an interest in the waters. Enforcement provisions deal with the following.

I Contraventions of sections 3(1) and 4(1). Local authorities, the Minister for Fisheries, Fisheries Boards and any private individual may prosecute in respect of breaches of sections 3(1) and 4(1) of the Act.[57] In addition or alternatively, section 10(1) of the Act empowers the aforesaid Minister, Fisheries Boards and local authorities to apply to the District Court for an order directing the person named therein to mitigate or remedy any effects of a contravention of sections 3 and 4 in such manner and within such time as may be specified in the court order. Alternatively, section 10(5) empowers a local authority (only) to serve a written notice requiring the ceasing of the contravention of sections 3(1) and 4(1) within such period as may be specified in the notice and requiring the mitigation or remedying of any effects of the contravention within the period and in the manner specified. In the event of non-compliance with a District Court order or on failure to comply with a notice under section 10(5), the local authority itself may take any steps it considers necessary to prevent the entry or discharge or to mitigate or remedy any effects of the contravention. It may also recover the cost of such steps from the person on whom the notice is served as a simple contract debt in a court of competent jurisdiction on satisfying the court that the person is responsible for the contravention. Non-compliance with a court order under section 10(1) is a criminal offence punishable on summary conviction by a fine not exceeding £250 together with a fine not exceeding £100 for every day the contravention is continued and/or 6 months' imprisonment.

Section 11 of the Act provides that a local authority or any person may apply to the High Court where a contravention of section 3(1) or 4(1) has occurred or is occurring and the Court may prohibit the continuance of the contravention and/or make such interim or interlocutory order or such order as to costs as it considers appropriate.

II Measures to prevent water pollution from premises. Section 12 empowers local authorities, in the interests of preventing or controlling water pollution, to deal with problems arising from the custody and storage of polluting matter on premises. Notice may be served on the person having custody or control of such matter, directing him to take specified measures considered necessary by the local authority to prevent polluting matter from entering waters. If the notice is not complied with as and within the time specified, the local authority may take any steps it considers necessary to prevent the polluting matter from entering waters at the expense of the person on whom the notice was served. Non-compliance with a notice is an offence punishable on summary conviction by a fine not exceeding £250 together with a fine not exceeding £100 for every day the offence is continued. A prosecution may be brought by a local authority.

III Powers to prevent and abate pollution in cases of urgency. Section 13 of the Act empowers local authorities to take such steps, carry out such operations or give such assistance as they consider necessary to prevent polluting matter entering waters, to remove such matter from the waters, to dispose of it as they think fit, and to mitigate or remedy the effects of any pollution caused by the matter. This power may only be exercised if, in the opinion of a local authority, 'urgent measures' are necessary to prevent pollution of any waters in its functional area, to remove polluting matter from waters in that area, or, while such matter is in waters outside that area, to prevent it from entering any part of that area. Expenses properly incurred by a local authority under section 13 are recoverable from the polluter.

This section gives a statutory basis for local authority involvement in clean-ups necessitated by oil spills and was partly motivated by this need. It can, of course, also be invoked in the event of other polluting substances entering waters or the sea. It is the only section in Irish legislation dealing—albeit obliquely—with civil liability for oil pollution.[58]

IV Accidental discharges. Local authorities may prosecute any person who does not comply with the obligation under section 14 of the Act to notify them as soon as practicable after the occurrence of an accidental discharge, spillage or deposit of any polluting matter which enters or is likely to enter any waters. The penalty on summary conviction is a fine not exceeding £250.

V Power to require information with respect to abstractions and discharges. Section 23 of the Act empowers a local authority to require the furnishing by any person who is abstracting water from or discharging effluent or other matter to any waters, of relevant information about the abstraction or discharge. Failure to give the required information within the time allowed or furnishing a written statement which is false or knowingly misleading in a material respect is an offence punishable on summary conviction by a fine not exceeding £100. The prosecution may be taken by a local authority.

VI Prosecutions. All offences under the Act relevant to discharges to waters may be prosecuted by local authorities. Offences under sections

3 and 4 may, in addition, be prosecuted by a Fisheries Board, the Minister for Fisheries, and a private individual. Offences under section 10(1) may be prosecuted by a Fisheries Board and the Minister for Fisheries. However, since bodies other than local authorities will not usually have access to monitoring records and do not have powers under section 23 to require information on discharges made or being made, it is probable that their prosecuting role will be somewhat limited and confined principally to cases where a pollution has actually occurred or where the person responsible for the pollution is a local authority. Section 31 of the Act empowers a local authority to prosecute for an offence whether or not the offence occurred in (or in respect of waters in) the authority's functional area.

VII Incidental powers. Section 28 of the Act provides the usual necessary powers of entry, inspection etc. to 'authorised' persons in relation to the performance of their functions under the Act. Of particular interest is the power of authorised persons to enter the premises (which expression includes land whether or not there are buildings thereon) at any time where urgent measures are necessary to prevent or abate pollution. Obstruction of an authorised person in the performance of his duty is punishable on summary conviction by a fine not exceeding £250 and/or 6 months' imprisonment. An 'authorised' person is a person appointed by a local authority, the Minister for Fisheries or a Fisheries Board.

4.4.7.1 EXTENT OF ENFORCEMENT

Although in force since 1977, the Water Pollution Act has not yet been implemented by many local authorities and sanitary authorities. By the end of 1979 less than a dozen licences had been issued under the Act although a large number of licence applications had been submitted. By the same date only four prosecutions had been brought under section 3. No other provisions in the Act have been judicially enforced. Delays in implementing the Act were caused by personnel difficulties in local authorities and by difficulties in setting up the infrastructure necessary.

4.5 DRINKING WATERS

Under the Public Health (Ireland) Act 1878, statutory responsibility for ensuring the provision of water fit for human consumption rests primarily with 87 sanitary authorities[59] which have a general power to supply their districts with 'a supply of water proper and sufficient for public and private purposes',[60] and which are specifically obliged to provide and keep in waterworks belonging to them a supply of 'pure and wholesome water'.[61] They may take proceedings to prevent pollution of waterworks within their jurisdiction from sewage[62] and may, on receipt of a complaint that potable water is so polluted as to be injurious or dangerous to health, seek a court order to eliminate such danger.[63] Where a sanitary authority is empowered to take water from any source, it has the same rights of preventing interference with the flow and pollution of the water as has the riparian owner.[64] Where a public water supply system has been provided, sanitary authorities may in certain circumstances require the connection therewith of any premises not provided with a satisfactory water supply.[65] Section 61 of the Waterworks Clauses Act 1847 prohibits the contamination of any stream or reservoir used as a public water supply, or any aquaduct or other part of the supply system. The penalty is £1 for each day that the offence continues. Section 17 of the Waterworks Clauses Act 1863 provides that it is an offence for any person wilfully or negligently to cause or suffer 'any pipe, valve, cock, cistern, bath, soil-pan, water closet, or other apparatus or receptacle to be out of repair, or to be so used or contrived' as to cause the water supplied by the sanitary authority to be wasted, misused, unduly consumed or contaminated or so as 'to occasion or allow the return of foul air, or other noisome or impure matter' into any pipe belonging to a sanitary authority. Every such offence is punishable by a fine not exceeding £5. Section 16 of the same Act permits the sanitary authority to cut off the water supply to anybody who fails in certain circumstances to prevent it being contaminated. These sections were incorporated into the Public Health (Ireland) Act 1878 by section 67 thereof.

The statutory powers of sanitary authorities in relation to, *inter alia*, drinking water sources have been greatly strengthened by the enactment of the Local Government (Water Pollution) Act 1977. This definition of 'polluting matter' in section 1 of that Act includes 'matter rendering waters harmful or detrimental to public health or to domestic . . . uses'.[66] This would include rendering drinking water unfit for such use. Of particular relevance are sections 3, 12, 13, 15, 22 and 23. The Department of the Environment in conjunction with the Geological Survey of Ireland (Groundwater Division) are developing an aquifer protection policy to ensure that, *inter alia*, groundwaters used extensively as a source of drinking water are not polluted.[67]

In fulfilling this statutory obligation to provide 'pure and wholesome water' sanitary authorities have generally complied with the recommendations of the World Health Organisation in the selection and treatment of drinking water sources. Since June 1977, however, they have been instructed by the Department of the Environment to comply with standards prescribed in the EEC Council Directive 75/440/EEC[68] concerning the quality required of surface water intended for abstraction of drinking water in the Member States.[69] The implementation of this Directive did not entail the enactment of any new legislation and was secured by the issue of circular letters to sanitary authorities.[70]

In March 1981 local authorities were instructed to implement EEC Council Directive 80/68/EEC on the protection of groundwater against pollution from certain dangerous substances.[70a] A further EEC Directive (80/778/EEC, as amended by 81/858/EEC) relating to the quality of water intended for human consumption must be fully complied with by 15 July 1985. Methods of measurement and frequencies of sampling and analysis of surface water intended for the abstraction of drinking water prescribed in EEC Directive 79/869/EEC, as amended by 81/855/EEC, must be complied with by October 1987. Legislation will not be necessary for the implementation of this Directive.

4.6 BATHING WATERS

Section 41 of the Local Government (Sanitary Services) Act 1964 provides that

> a sanitary authority may make by-laws for the regulation of public bathing within their sanitary district, and the by-laws may, in particular, provide for all or any of the matters mentioned in the Third Schedule to this Act.

Matters mentioned in the Third Schedule include the prohibition of public bathing except in specified areas. Some sanitary authorities have made by-laws under section 41. None, however, has made any requirements as to the quality of waters suitable for bathing.

EEC Council Directive 76/160/EEC concerning the quality of bathing water[71] was implemented administratively in Ireland partly by circular letter to sanitary authorities[72] and partly by the *Memorandum on Water Quality Guidelines*[73] issued by the Department of the Environment to all local authorities. The latter contains recommendations on the values applicable to bathing waters for the parameters in the Annex to the Directive. Claims that the Directive has been implemented have been challenged on a number of occasions. The EEC definition of 'bathing waters' is 'waters in which:

(a) bathing is expressly authorised by the competent authorities; or

(b) bathing is not prohibited and is traditionally practised by a large number of bathers'.[74]

Bathing is not 'expressly authorised' in Ireland so that the sole criterion for identifying a bathing place is (b) above. On at least two occasions[75] the Minister for the Environment has denied in the Dail that specific locations were bathing places on the grounds that they were not used by 'a large number of bathers'. His perception of a large number is apparently '10,000 persons per linear kilometre'—a figure which, as he himself admitted, is not exceeded even 'at the most popular beaches in Ireland'. One must therefore conclude from the Minister's reported statements that there are no 'bathing waters' within the meaning of the Directive in Ireland. However, in 1981 it was decided that the following bathing places would be required to comply with the spirit of the Directive:

Portmarnock and Dollymount, Co. Dublin;

Courtown, Co. Wexford;

Fountainstown, Co. Cork;

Salthill, Co. Galway;

Tramore, Co. Waterford.

National limit values have been set for waters in these areas.[76]

4.7 OFFENSIVE TRADES

It is an offence under section 128 of the Public Health (Ireland) Act 1878 to establish within the district of any urban authority, without written consent, any offensive trade. The following trades are offensive trades for the purposes of the Act: the trade of blood-boiler, bone-boiler, fellmonger, soap-boiler, tallow-melter, tripe-boiler, gut manufacturer and any other noxious or offensive trade, business or manufacture *ejusdem generis* with those specified. The penalty is £50 plus £2 for each day on which the offence is continued. Urban authorities are empowered to declare a business an offensive business[77] and to make by-laws with regard to any trade which is an offensive trade in order to prevent or diminish any noxious or injurious effects therefrom.[78] Where an urban authority medical officer or two medical practitioners or ten inhabitants of an urban district certify that an offensive trade is a nuisance or injurious to health, an urban authority must bring proceedings against that trade. It is a defence for the offender to prove that he has used the best practicable and available means for abating such nuisance or for preventing or counteracting any effluvia which is a nuisance or injurious to health. The maximum penalty on conviction is £5 on the first instance, but further penalties up to £200 may be imposed on subsequent convictions.[79]

4.8 AGRICULTURAL EFFLUENTS

The significance of agricultural effluents as the third most important source of pollution merits treating them under a separate heading.

The use of land for the purposes of agriculture and forestry is exempted development under section 4 of the Local Government (Planning and Development) Act 1963. Accordingly, as the Inter-departmental Environment Committee pointed out,

> the Act does not offer appropriate means to deal with pollution or other effects detrimental to the environment resulting from fertilizer run-off, chemical seed dressings, herbicides and insecticides, and silage-making operations.[80]

However, the development of buildings for housing livestock and poultry exceeding 400 square metres (whether or not by extension of an existing structure) and any ancillary provision for effluent storage has been subject to planning control since 15 March 1977.[81]

The entry of agricultural effluents into waters may be an offence under section 3(1) of the Local Government (Water Pollution) Act 1977 but, unless they enter waters or a sewer via 'any works, apparatus, plant or drainage pipe' used for their disposal to waters or a sewer, it would appear that discharges or entries of agricultural effluents to waters are not licensable under sections 4 and 16 of the Act. The Technical Committee on Effluent and Water Quality Standards has recommended treatment methods for animal manures but has stated that 'at the present time, it is not economically practicable to treat animal manures or silage effluent to a degree which would allow discharge to waters under licence'. In the Committee's opinion 'it is not anticipated that the question of issuing licences to discharge such effluents will arise'.[82]

In other respects, however, provided they come within the definition of 'polluting matter' in section 1 of the Water Pollution Act, agricultural effluents must be treated identically to other polluting effluents. In practice, it is likely that preventative legal control over pollution by agricultural effluents (other than run-off) is most likely to be exercised under section 12 of the Act, which empowers local authorities to require specified steps to be taken to prevent polluting matter entering waters from premises (e.g. silos, livestock housing, slurry tanks, dungsteads).[83]

The reality of environmental control over pollution by agricultural activities, however, is that the primary and probably the most effective controls are extra-legal. Farmers are eligible for substantial grant-aids from the Department of Agriculture when carrying out agricultural developments. The scheme under which these grant-aids are administered is known as the Farm Modernisation Scheme. Approvals for grant-aids will only be given where planning permission has been granted for the proposed development or where the conditions for exemption from the necessity to apply for planning permission are clearly satisfied. In issuing approvals for grant aids, account is taken of the necessity for pollution control, and conditions are attached, where necessary, to safeguard against the possibility of pollution from animal wastes. Guidelines on appropriate conditions are contained in a book entitled *Guidelines and Recommendations on Control of Pollution from Farm Wastes*[84] circulated by the Department of Agriculture, the purpose of which is

> to set out a basis for the application of uniform guidelines and recommendations on pollution control pertaining to animal buildings and other waste-producing structures on the farm and to suggest appropriate minimum standards or conditions on which payment of grant-aid should, for the future, be offered under the Farm Modernisation Scheme in respect of such structures.

The guidelines are flexible and provision is made for deviation from them where the circumstances pertaining to any particular development justify stricter or more lenient controls. There is, however, nothing to prevent a sufficiently motivated farmer from avoiding any constraints imposed by the administrators of the Farm Modernisation Scheme provided he is willing to forego the grant-aid.

Further advice on all aspects of pollution control relevant to farming activities is available to farmers through the Agricultural Advisory Services and An Comhairle Oiliuna Talmhaiochta.

Persons responsible for pollution caused by agricultural activities may also be liable to criminal penalties under the various statutes penalising water pollution described in this section.

4.9 WATERBORNE CRAFT

Controls over pollution of inland waters by waterborne craft are exercisable under the Local Government (Water Pollution) Act 1977, section 27 of which empowers the Minister for the Environment, after consultation with the Minister for Transport, the Commissioners of Public Works in Ireland and the Water Pollution Advisory Council, to make regulations enabling local authorities

> to prohibit, restrict or regulate the keeping or use in such waters (other than tidal waters) as may be specified in the regulations, of

vessels with sanitary appliances from which polluting matter passes or can pass into the waters.

Sanitary authorities are empowered to provide facilities for the reception and disposal of sewage from vessels and to impose fees or other charges for the use of such facilities. Contravention of any regulation is an offence punishable on summary conviction by a fine not exceeding £250 together with a further fine of £100 for every day on which the contravention is continued. Prosecutions under section 27 may be taken by a local authority. Regulations have not yet been made under section 27. Causing or permitting the entry of polluting matter from waterborne craft to waters other than tidal waters is an offence under section 3(1) of the Act and may also be an offence under various other statutes penalising pollution described in this section.

4.10 MISCELLANEOUS

A number of provisions in various statutes of little general importance from the point of view of water pollution control also prescribe penalties for causing water pollution. These include section 47 of the Public Health Acts Amendment Act 1890, which prohibits the throwing of cinders, ashes, bricks, stones, rubbish, dust, filth or other matter likely to cause annoyance into any river, stream or watercourse. The penalty is a fine not exceeding £2. Section 4 of the Alkali etc. Works Regulation Act 1906 prohibits the discharge of alkali waste into waters or otherwise 'without the best practicable means being used for effectually preventing any nuisance from arising therefrom'. The penalty is £20 for a first offence and £50 plus £5 for each day a subsequent offence continues. Section 42 of the Electricity (Supply) (Amendment) Act 1945 prohibits any person from discharging or allowing to escape into a river or stream serving electricity generating stations any chemical or other substance likely to injure part of the generating station or any subsidiary or connected works without the written permission of the Electricity Supply Board and compliance with conditions attached to such permission. Contravention entails a £50 fine, plus £20 for each day on which the offence is continued.

4.11 INDIVIDUAL RIGHTS

At common law, an individual may have rights to sue in negligence, nuisance, trespass or under the rule in *Rylands* v. *Fletcher* for injury to his interests caused by water pollution. There are, however, limitations on the usefulness of these actions as anti-water-pollution devices,[85] and the common law itself has developed specific remedies whereby riparian owners (i.e. the owners of land in actual and reasonably proximate contact with a watercourse) have specific rights to ensure that water is not polluted.[86] The riparian owner on a natural watercourse flowing in a known and defined channel, either subterranean or on the surface, has a proprietary right to have the water flow past his land 'without sensible alteration in its character or quality'.[87] Therefore, if waters which a riparian owner is entitled to use are polluted, he may sue the polluter. At common law, pollution means adding anything to water which changes its natural qualities: the expression includes raising the temperature of water,[88] adding hard water to a soft water stream,[89] and discharging sewage and refuse from a factory.[90] Proof of actual damage is not required: it is sufficient to prove that the right to receive water in its natural state has been infringed.[91] The rights of riparian owners do not apply where underground water flows in a defined but unknown channel or where water merely percolates through the soil, but an action for nuisance may lie instead. In one case a landowner was successfully sued for nuisance when he polluted his land so that water percolating from it to his neighbour's land was polluted.[92] The owner or grantee of fishing rights may also sue in respect of injury to his rights caused by water pollution.[93]

There are also a number of statutory remedies which the individual can assert in respect of water pollution. Under the Public Health (Ireland) Act 1878, an individual may require his sanitary authority to take action against a statutory nuisance or he himself may initiate a prosecution for such a nuisance.[94] Any individual may seek an order under section 27 of the Local Government (Planning and Development) Act 1976 prohibiting an unauthorised development causing water pollution or enforcing water pollution control conditions attached to a planning permission.[95] Any individual may prosecute, under sections 3 and 4 of the Local Government (Water Pollution) Act 1977, a person who causes or permits the entry of polluting matter to waters or who discharges trade or sewage effluents to waters other than under or in accordance with a licence granted under section 4 of the same Act.[96] In addition, under section 11 of the Water Pollution Act, any person, whether or not he has an interest in the waters, may seek a High Court order prohibiting the continuance of a contravention of sections 3(1) and 4(1) of the Act.

4.12 MONITORING

The Local Government (Water Pollution) Act 1977 contains extensive and wide-ranging provisions for the monitoring of all waters (inland or sea) and discharges of effluents to waters. The following table, adapted from a *Report on Monitoring Industrial Pollution*, published by the Institute for Industrial Research and Standards,[97] is a summary of the monitoring provisions in the Act.

Section	Purpose of monitoring	Form of monitoring required	Monitoring agency
3	To ensure compliance with the general prohibition on the entry of polluting matter other than trade or sewage effluent to waters.	Inspection, investigation, or sampling to determine if an unauthorised discharge has taken place, or is taking place.	LA FB MF AN Note: AN may not enter a premises for monitoring purposes unless authorised by SA, LA, FB, MF or DOE.
4, 16	To ensure compliance with the conditions attached to a licence to discharge effluent.	(a) Measurement of the nature, composition, temperature, volume rate, and time of discharge. (b) Inspection of the method of treatment and location of discharge. (c) Monitoring the effect of a discharge on the receiving waters. (d) Checking on the design and construction of outlets for a discharge. (e) Providing or checking on the provision and maintenance of meters, gauges, other apparatus, manholes and inspection chambers. (f) Taking and analysis of samples, keeping of records, and providing information to the local authority.	D LA SA FB MF AN Note 1: AN may not enter a premises for monitoring purposes without authorisation as noted above. Note 2: An industrialist may be required by licence conditions to pay for or contribute towards the cost incurred by a Local Authority in monitoring a discharge.

Section	Purpose of monitoring	Form of monitoring required	Monitoring agency
22	For the LA or SA to carry out any of its functions under the Act (general monitoring provision).	(a) Any monitoring, sampling, measurement or analysis of waters, effluents and other matter, which the LA or SA considers necessary. (b) Collecting any information the LA or SA considers necessary for performing its functions under the Act. (c) Providing meters, gauges, manholes or any other apparatus for any of the purposes of this section.	
23	For any purpose relating to the functions of an LA or SA under the Act.	Obtaining and supplying in writing to the LA or SA any details regarding an abstraction or discharge sought in a notice from the Authority.	D LA SA
26	In the case of sewage effluents or waters to which such effluents discharge, to ensure that the discharge complies with any relevant water or effluent standard prescribed by the Minister of the Environment under this section.	Sampling and analysis of sewage effluents and receiving waters.	SA
27	To enable the LA to prohibit, restrict, or regulate water pollution by sewage from boats.	Inspection of sewage disposal systems in boats in non-tidal waters.	LA
6, 19	To provide specified information or evidence to verify any information given by the applicant for a licence.	A special or short-term survey or investigation of a discharge, probably involving some flow measurement, sampling and analysis.	D

Section	Purpose of monitoring	Form of monitoring required	Monitoring agency
7, 17	To provide information required in connection with a licence review, e.g. to establish whether the discharge is a significant threat to public health, or if a material change has taken place in the receiving water.	(a) Investigation of whether an effluent is a threat to public health: possibly toxicity tests. (b) Review of receiving water conditions to determine if there is any recent significant change.	D LA
12	To determine whether a person has the custody or control of any polluting matter which should be prevented from entering waters.	Inspection of premises for matter which may pose a risk of water pollution.	LA SA
14	To identify and control accidental discharges.	Checking for occurrence of accidental discharges.	D LA SA
16	To ensure polluting water is not discharged to surface water or storm drains.	(a) Checking layout of foul sewers and surface water drains. (b) Sampling surface or storm water drains to ensure no polluting matter has been or is being discharged.	D SA
28	(a) To perform any function conferred under this Act on LA, SA, DOE, MF or BC. (b) To find out whether such a function should be performed. (c) To find out whether the Act is being or has been contravened in any way.	Carrying out any necessary inspection and taking any necessary samples in premises or vessels.	LA SA DOE MF FB AN Note: AN may enter premises for monitoring purposes only if authorised by LA, SA, DOE, AF or FB.

Abbreviations: D, discharger; LA, local authority; SA, sanitary authority; FB, Fisheries Board; MF, Minister for Fisheries; DOE, Department of the Environment; AN, any other person.

One feature which becomes obvious from the table is the multiplicity of authorities empowered to monitor the same effluent. Under section 24 of the Act the Minister for the Environment may make regulations requiring local authorities, sanitary authorities and Fisheries Boards 'to consult with such persons and in such manner in relation to the exercise of such powers and duties under this Act as may be prescribed'. The Minister has not yet made any regulations under section 24 but all public authorities involved in water pollution control have been repeatedly urged to cooperate with each other and to coordinate their activities. Few public authorities have the capacity (nor indeed would it be administratively or economically feasible for them) to carry out their monitoring powers and duties under the Act. It is likely that, at least until Water Quality Control Authorities are established under section 25, the vast bulk of monitoring will be carried out on behalf of control authorities by An Foras Forbartha or the Institute for Industrial Research and Standards. In 1976, the Institute calculated that if all 'significant' (determined by reference to their effect on receiving waters) industrial discharges were monitored, over 497 factories would be involved and monitoring costs would be about £3¼ million (£1 million for equipment and £2¼ million for testing and analysis).[98] It is extremely unlikely that local authorities will embark on monitoring all these discharges and it is generally believed that primary responsibility for monitoring industrial discharges will remain with the discharger, with local authorities carrying out periodic inspections of monitoring equipment and occasional checks on monitoring results.

Apart from monitoring obligations necessitated by the implementation of the Water Pollution Act, control authorities are obliged to monitor drinking and bathing waters in order to comply with EEC Directives already referred to. Methods of sampling and analysis if prescribed by Directives must be adopted. Monitoring is also required for the proper implementation of EEC Directives on pollution caused by certain dangerous substances discharged into the aquatic environment of the Community and on the quality of fresh waters needing protection or improvement in order to support fish life. A number of rivers have been designated by circular letters as salmonid waters in need of protection or improvement in accordance with the requirements of the last mentioned Directive.[99] Bathing waters are monitored at six locations around the coast. An Foras Forbartha also monitors the quality of surface waters at four locations in accordance with the provisions of the EEC Decision establishing a common procedure for the exchange of information on the quality of surface water in the Community.[100] Other bodies monitoring discharges to waters are shown in the following table:[101]

Agency	Monitoring function
Department of Agriculture	Monitors pesticide levels in water
Agricultural Institute	Carries out various monitoring projects for research purposes
Department of Fisheries and Forestry	Monitors fishery waters and suspected pollution sources including licensed discharges
Fisheries Boards	Monitoring fishery waters
Inland Fisheries Trust	Monitoring fishery waters
Foyle Fisheries Commission	Monitoring fishery waters
Eastern and Western Health Boards	Monitoring bathing and drinking waters
Institute for Industrial Research and Standards	Baseline monitoring in relation to specific industrial projects
	Monitoring commissions for public and private sector clients
Local Authorities	Monitoring for observance of planning control conditions
	Monitoring of water pollution and measurement of river flows
An Foras Forbartha	On-going water quality survey in rivers and lakes
	Monitoring surface fresh waters
	General monitoring service to local authorities
Office of Public Works	River flow measurement in connection with arterial drainage
Bord Failte	Monitoring development proposals of particular importance for tourism
Coras Iompar Eireann	Laboratory examination of pollution occurrences in canals
Electricity Supply Board	Monitoring of waters in their charge

4.13 WATER QUALITY STANDARDS

At present, the only mandatory water quality standards in Ireland are those prescribed in the EEC Directives implemented here and referred to in this chapter. Irish legislation does not specify standards for particular emissions or for general water quality. Each control authority established by legislation on water pollution has a discretion to set its own standards. When setting standards, authorities have regard among other things to local conditions and circumstances, to the current state of scientific and technical knowledge, and to the financial implications for the discharger and, where appropriate, the authority itself.

Section 26 of the Local Government (Water Pollution) Act 1977 empowers the Minister for the Environment to make regulations prescribing:

(i)   quality standards for water and sewage effluents;

(ii)   standards for methods of treatment of such effluents.

The Minister has not made any regulations under this section. Instead, a Technical Committee of Effluent and Water Quality Standards appointed to advise the Minister on water quality and emission standards has issued a *Memorandum No. 1 on Water Quality Guidelines* 'to provide local authorities with guidelines on water and effluent quality for their assistance in dealing with development proposals which may affect water quality'. The guidelines are confined to recommending quality objectives for the protection of fisheries and of man insofar as he consumes water or the fishery resources therein.[102] The Committee has shown a general preference for the environmental quality objectives approach to water quality management over the uniform emission standards approach. The former approach allows the discharge of pollutants to waters provided stated quality objectives for the waters are not infringed; the latter requires uniform emission standards for all discharges regardless of the assimilative capacity of the receiving water. However, where highly toxic and persistent pollutants listed in List 1 of the EEC Directive 76/464/EEC (on pollution caused by certain dangerous substances discharged into the aquatic environment of the Community) are concerned, the Committee recommends that emission standards be adopted. These recommendations are of course merely guidelines and have no statutory force. It is anticipated that the Committee will issue further memoranda at future dates. Various circular letters from the Minister of the Environment acquaint local authorities of the quality and emission standards required under EEC Directives and advise them to ensure compliance with the Directives in the appropriate manner. This is usually by the attachment of conditions to licences under the Water Pollution Act and for permissions under the Local Government Planning and Development Acts 1963–1976.

There is little published information on the standards or guidelines actually applied by control authorities. Where industrial developments are concerned, however, standards recommended by the IIRS and/or An Foras Forbartha are most likely to appear in conditions attached to planning permissions, grant-aids and licences under the Water Pollution Act. In general, IIRS claims to abide by any EEC standards whether these have become mandatory or not. Otherwise standards set by such bodies as the United States Environmental Protection Agency or the United States Public Health Service or the European Inland Fisheries Advisory Commission are referred to.[104]

Guidelines on sampling and analysis of waters and effluents were circulated by the Department of the Environment to local authorities in February 1981.[105]

Quality standards contained in Directive 78/659/EEC on the quality of fresh waters needing protection or improvement in order to support fish life have been adopted as national standards pending the prescription of standards by the Minister for the Environment. These standards apply in respect of:

the rivers Moy and tributaries except the Tubbercurry River, Blackwater, Lough and River Corrib, Fergus, Feale, Swilly, Finn, Nore, Slaney, Lee (above Cork City waterworks), Boyne, Dargle, Vartry, Aherlow, Bride, Arigideen, Brown, Flesk, Maine, Lurgay, Glashagh, Lennon and Maggisburn.

They have been designated as being in need of protection and improvement in accordance with the provisions of the Directive. Their waters must comply with the prescribed standards within five years of designation.[106]

The most recent publication on Irish river water quality is the *National Survey of Irish Rivers: A Review of Biological Monitoring 1971–79*, published by An Foras Forbartha (1981). This report, *inter alia*, indicates trends in water quality in a representative selection of Irish rivers.

Notes

1. Annual Report of An Foras Forbartha (1979). See also *National Survey of Irish Rivers: A Review of Biological Monitoring 1971–79*, An Foras Forbartha (1981).
2. Public Health (Ireland) Act 1879, s. 108.
3. *Ibid.*, ss. 109–112, 113, 123.
4. *Ibid.*, s. 121.
5. Fisheries (Amendment) Act 1962, s. 2(1). Fisheries Act 1980, s. 50.
6. See 2.4.4.
7. Local Government (Water Pollution) Act 1977, s. 34(c).
8. Fisheries (Consolidation) Act 1959, ss. 309, 312. Fisheries Act 1980, s. 50.
9. *Dail Reports*, Vol. 312, Col. 623, 1 March 1979.
10. By-Law No. 589, 1976.
11. See Chapter 2.
12. See 2.4.4.
13. Local Government (Water Pollution) Act 1977 (Commencement) Order 1977 (S.I. No. 117 of 1977).
14. Local Government (Water Pollution) Act 1977 (Fixing of Dates) Order 1978 (S.I. No. 16 of 1978).
15. Local Government (Water Pollution) Act 1977, s. 1.
16. See 4.3.
17. *Alphacell* v. *Woodward* [1972] A.C. 824.
18. *Price* v. *Cormack* [1975] 2 A11 E.R. 113.
19. Local Government (Water Pollution) Act 1977, s. 3(3).

20. See Manion, L., *Guidelines and Recommendations on the Control of Pollution from Farm Wastes* (1977), Department of Agriculture.
21. *Ibid.*, s. 3(5).
22. *Ibid.*, ss. 3(4), 3(2).
23. *Ibid.*, s. 1.
24. *Ibid.*, s. 26(3).
25. Local Government (Water Pollution) Regulations 1978 (S.I. No. 108 of 1978).
26. *Ibid.*, art. 4 and First Schedule.
27. See 2.4.2 and 2.4.3.
28. Local Government (Water Pollution) Regulations 1978, art. 5.
29. *Ibid.*, art. 6. See 2.4.2.
30. *Ibid.*, arts. 6(3) and 11(1)(*d*).
31. *Ibid.*, art. 7.
32. *Ibid.*, arts 7(3), 8, 9.
33. *Ibid.*, art. 10.
34. Local Government (Water Pollution) Act 1977, s. 4(3).
35. See *Memorandum No. 1 Water Quality Guidelines* (1979), Stationery Office.
36. Local Government (Water Pollution) Act 1977, s. 4(5).
37. Circular ENV. 9/78 of 7 April 1978.
38. Circular ENV. 2/80 of 24 April 1980.
39. Local Government (Water Pollution) Regulations 1978, art. 11.
40. Local Government (Water Pollution) Act 1977, s. 9.
41. *Ibid.*, s. 4(6).
42. *Ibid.*, s. 4(7).
43. *Ibid.*, s. 4(9).
44. *Ibid.*, s. 4(11).
45. See 4.4.2.
46. Local Government (Water Pollution) Regulations 1978, art. 13.
47. *Ibid.*, art. 14(1) and (2).
48. *Ibid.*, art. 14(3).
49. Local Government (Water Pollution) Act 1977, s. 7(2).
50. Local Government (Water Pollution) Regulations 1978, art. 15.
51. Local Government (Water Pollution) Act 1977, s. 8. Local Government (Water Pollution) Act 1977 (Transfer of Appeals) Order 1978 (S.I. No. 96 of 1978).
52. Local Government (Water Pollution) Regulations 1978, art. 26.
53. See 2.4.5.
54. Local Government (Water Pollution) Regulations 1978, art. 34.
55. *Ibid.*, art. 35.
56. Local Government (Water Pollution) Act 1977, ss. 4(3) and 16(2).
57. See 4.4.2 and 4.4.3.
58. See 6.2.
59. Public Health (Ireland) Act 1878, s. 74.
60. *Ibid.*, s. 61.
61. *Ibid.*, s. 65.
62. *Ibid.*, s. 78.
63. *Ibid.*, s. 79.
64. Water Supplies Act 1942, s. 20. See also 4.11.
65. Local Government (Sanitary Services) Act 1972, s. 8(2).
66. See 4.4.2 and 4.4.7.
67. See Wright, G. R. and Daly, D., 'The Hydrogeological Aspects of Tip Site Selection'. Paper at Conference on Planning for Waste Disposal (1979), An Foras Forbartha.
68. OJ L 194, 25 July 1975.
69. Circular L 1/77 of 17 June 1977.
70. Circulars L 4/75 of 20 December 1975; L 5/75 of 2 December 1975; L 1/77 of 17 June 1977.

70a. Circular ENV 3/1981 of 25 March 1981.
71. OJ L 31, 5 February 1976.
72. Circular ENV. 14/76 of 28 July 1976.
73. See 4.13.
74. EEC Council Directive 76/160/EEC concerning the quality of bathing water, art. 2.
75. *Dail Reports*, Vol. 310, Col. 478, 30 November 1978; Vol. 320, Col. 1414, 13 May 1980.
76. Circular ENV 3/1981 of 22 May 1980.
77. Public Health Acts Amendment Act 1907, s. 51.
78. Public Health (Ireland) Act 1878, s. 129 as amended by Public Health Acts Amendment Act 1907, s. 51.
79. *Ibid.*, s. 130.
80. *Report on Pollution Control*, p. 5.
81. Local Government (Planning and Development) Regulations 1977, Third Schedule, Part III, Class 7.
82. *Memorandum on Water Quality Guidelines*, p. 48.
83. See 4.4.7.
84. Mannion, L., *Guidelines and Recommendations on Control of Pollution from Farm Wastes* (1977), Department of Agriculture.
85. See 1.2.
86. See Gale, *Easements* (14th edn), pp. 229–234.
87. *Young v. Bankier Distillery Co.* [1893] A.C. 698.
88. *Tipping v. Eckersley* [1855] 2 K & J 264.
89. *Young v. Bankier Distillery Co.* [1893] A.C. 698.
90. *Crossley & Sons Ltd. v. Lightowler* (1867) 2 Ch. App. 478.
91. *Ibid.*
92. *Ballard v. Tomlinson* (1885) 29 Ch. D. 126.
93. *Fitzgerald v. Firbank* [1897] 2 Ch. 96.
94. See 4.1.
95. See 2.4.8.
96. See 4.5.2 and 4.5.3.
97. Lynch, M., 'The Monitoring Requirements of the Local Government (Water Pollution) Bill, 1976', in *Monitoring Industrial Pollution* (1976), IIRS, pp. 35–39.
98. *Monitoring Industrial Pollution*, p. 50.
99. Circulars ENV 9/1980 of 25 September 1980, ENV 14/82 of 14 September 1982.
100. OJ L/334, 24 December 1977. See Circular ENV. 4/78 of 24 February 1978.
101. *Report on Pollution Control*, pp. 52–56.
102. *Memorandum No. 1 on Water Quality Guidelines*, p. 19.
103. *Ibid.*, p. 17.
104. See Lynch, M., 'Chemical Aspects of Water Pollution', Seminar on Effluents (1980), Cork Scientific Council.
105. See Circular ENV 9/1980 of 25 September 1980 and DOE Letter of 13 November 1981 to affected County Councils. The first seven rivers were designated in July 1980 and the remainder in November 1981.
106. See Circulars ENV 1/1981 of 27 February 1981, ENV 14/82 of 14 September 1982.

5
Pollution of Coastal Waters

Ireland's coastal waters, running along 1738 miles of coastline, are, except for areas adjacent to major cities and industrial centres, relatively unpolluted. In recent years, however, the coastal zone has been subject to increasingly strong pressures from almost every activity. Over fifty per cent of the population now live by the coast and a great number of industries are located in coastal areas. Further urban growth, the concentration of major industries in coastal areas, the offshore dumping of wastes, and the development of offshore oil and gas industries are all likely to result in increasing volumes of wastes entering coastal waters.

Controls over pollution of coastal waters are exercised almost exclusively by public authorities acting under statutory powers. These authorities are the Ministers for Transport, Energy, Fisheries and Environment, local authorities and harbour authorities.

Apart from legislation controlling oil pollution, which is dealt with in the next section, the only specific legislation which could be used to control pollution of coastal areas (including the sea) is the Foreshore Act 1933, which confers wide control powers on the Minister for Transport. Legislation for the control of pollution of inland waters described in the preceding section usually applies also to coastal waters although there may be consequent differences in its practical administration and enforcement. In this section it is proposed to deal with legislation specifically applicable to coastal waters and to indicate the extent to which legislation described in the preceding chapter of this book applies to coastal waters.

5.1 FORESHORE ACT 1933

The State owns almost all of the Irish foreshore. In addition, the Foreshore Act 1933 empowers the Minister for Transport to purchase or lease by agreement non-State-owned foreshore.[1] For the purposes of the Act the word 'foreshore' means the bed and the shore below the line of high water of ordinary or medium tides of the sea and of every tidal river and tidal estuary and of every channel, creek, and bay of the sea or of any such river or estuary.[2] The Act establishes a system under which the Minister for Transport may exercise a certain degree of control over development on or near foreshores and distinguishes between his powers in relation to (a) State-owned foreshore and (b) other foreshore.

5.1.1 State foreshore

Section 2 of the Act enables the Minister to grant a lease of State foreshore if it is in the public interest to do so. The lease may refer to the foreshore itself and to any buildings or structures thereon. It may also include any minerals to a maximum depth of 30 feet from the surface of the foreshore together with a right to exploit those minerals. Leases must contain such terms as the Minister shall consider proper or desirable in the public interest and shall agree with the lessee, and a power or proviso for re-entry for breach, non-performance or non-observance of any term thereof. The Minister has a discretion to hold a public inquiry in regard to the making of a lease under section 2.

Section 3 of the Act empowers the Minister, if it is in the public interest, to grant a licence of State foreshore which authorises the licensee to place or erect any articles, things, structures, or works in or on such foreshore, to remove any beach material from such foreshore for any purpose. Every licence must contain such terms as the Minister shall consider proper or desirable and shall agree with the licensee. The Minister has a discretion to hold a public local inquiry before deciding to grant a licence under section 3.

There are approximately 150 leases and licences in force under the Act. Minerals exploitable under these leases and licences may include all minerals within the meaning of the Minerals Development Act 1940, other than scheduled minerals, mineral compounds and mineral substances.[3]

The Minister for Transport may thus control developments and other activities on State foreshore. Conditions relating to pollution control could in theory be included as terms in leases and licences granted under the Act. In practice, however, the Minister sees his functions as being confined to ensuring the safety of navigation and fisheries. Occasionally, terms in leases and licences have been used, directly or indirectly, as pollution controls. An example of the use of a foreshore lease to control pollution is the controversial lease granted to Gulf Oil Terminals (Ireland) Ltd. for the development of an oil terminal in Bantry Bay. That particular lease contained pollution control conditions and a proviso stipulating that in the event of pollution control arrangements proving inadequate or defective, the Minister would have the right to set up a harbour authority in the area[4]—a right which he eventually exercised.[5] Many licences are granted to sanitary authorities authorising them to place sewage disposal pipes in or on the foreshore or to industrialists disposing of effluent to coastal waters. Pollution control conditions are not attached to these licences.

Section 12 of the Act provides that the erection of any building, pier, wall or other structure on State foreshore must be authorised by the Minister. Where such erection is carried out without lawful authority the Minister may obtain a court order requiring or permitting that it be pulled down or removed within a specified time. Section 8 of the Act empowers the Minister to make regulations in respect of the public use of State foreshore. The Minister has not made any regulations under this section.

5.1.2 Non-State foreshore

Controls exercisable over privately-owned foreshore are naturally less extensive. Section 9 of the Act provides that the erection of sea defence works on such foreshore by any person other than its owner must be authorised by Ministerial order and carried out in accordance with the conditions and restrictions in the order. Section 10 provides that the erection of any building, pier, wall or other permanent structure must be carried out in accordance with maps, plans and specifications approved by the Minister but that the Minister may only withhold his approval on the grounds that the proposed structure would obstruct navigation or fishing.

5.1.3 Deposit of materials on foreshore, seashore or tidal lands

Section 13 of the Act requires that Ministerial consent be obtained and the conditions therein observed for the deposit of any material whatsoever on any foreshore or seashore or on any other place from which such material would escape by the operation of natural causes or be transplanted to such foreshore. The penalty on summary conviction is a fine not exceeding £10 but the Court may also order the convicted person to remove the materials within a specified time. Contravention of the court order is punishable by a maximum fine of £10 plus £1 for every day the contravention continues. The 'seashore' means the foreshore and every beach, bank and cliff contiguous thereto and all sands and rocks.[6]

Section 14(1) prohibits the throwing, depositing, leaving on any tidal lands or throwing into the sea adjacent to such lands of any glass, china, earthenware or other article which would injure a person bathing or wading on or from such lands or otherwise using such lands; the same prohibition applies to any material or substance (whether solid or liquid) which would or might be injurious or offensive to any such person. The penalty on summary conviction is a fine not exceeding £5. The expression 'tidal lands' means the bed and shore below the line of high water of ordinary spring tides of the sea and of every tidal river and tidal estuary and of every channel, creek and bay of the sea or of any such river or estuary.[7]

5.1.4 Removal of beach material from the seashore and foreshore

Under section 6 of the Act the Minister may make an order prohibiting the removal by any person of beach material of any kind or of any particular kind or kinds from the seashore whenever he is of the opinion that the removal of such material has prejudicially affected any public rights in respect of the seashore or any lands or waters in the neighbourhood thereof or had caused or is likely to cause injury to any land or to any building, wall, pier, or other structure. The Minister has a discretion to revoke or amend any prohibitory order and he may, if he thinks fit, hold a public inquiry into the continuation, making, amendment or revocation of such order. Contravention of a prohibitory order is punishable by a maximum fine of £10 and forfeiture of the beach material removed. 'Beach material' means sand, clay, gravel, shingle, stones, rocks and mineral substances on the surface of the seashore and includes outcrops of rock or any mineral substance above the surface of the seashore and seaweed on the seashore.[8]

Under section 7 of the Act the Minister may serve a notice prohibiting any person from removing beach material from State foreshore whenever he is of the opinion that its removal should be restricted or controlled. Contravention of the order is punishable by a maximum fine of £5 for a first offence and £10 for subsequent offences.

5.1.5 Enforcement

The Act is enforceable by the Minister for Transport. Enforcement of leases granted under section 2 may be by:

(i)   re-entry on the leased foreshore for breach, non-performance or non-observance of any covenant, condition or agreement;

(ii)   action for breach of a term in a lease;

(iii) threat of non-renewal if the lessee appears to require an extension of a lease.

Licences granted under section 3 may be enforced by:

(i)   termination;

(ii)   action for breach of a term;

(iii) threat of non-renewal.

The Minister has not taken any prosecutions under the Act in recent years.

5.2 HARBOURS ACT 1946

Harbour authorities have jurisdiction over harbour areas, the limits of which are indicated in the Harbours Act 1946.[9] A harbour authority is obliged, *inter alia*, to take proper measures for the management, control and operation of its harbour,[10] for the maintenance of all property and facilities under its control[11] and for the cleaning, scouring, deepening, improving and dredging of its harbour and approaches thereto.[12] It has power to provide facilities at its harbour including storage facilities, ballast and oil reception facilities;[13] to remove obstructions within the limits of the harbour;[14] and to place and maintain buoys.[15] Harbour authorities have a general power to make by-laws 'for the good rule and government of their harbours' including by-laws 'providing that the harbour master may remove nuisances from within the limits of the harbour'.[16] By-laws made by harbour authorities are not generally motivated by pollution control considerations except perhaps where oil pollution is concerned, and then only to a limited extent.

Harbour masters must be notified of the arrival of vessels at harbours[17] and may, subject to harbour by-laws, give directions to the masters of vessels for certain purposes, including the protection of persons or property and the regulation of traffic.[18] Hazardous goods brought within harbour limits must be properly and distinctly marked and the harbour authority may prohibit the bringing of dangerous articles within harbour limits or any specified parts of such limits.[19] It is an offence punishable by a maximum fine of £10 to put, cause or allow to be put, ballast, earth, ashes, stones or any other substance or thing into harbour waters without authorisation from the harbour authority.[20] All of the above provisions may operate directly or indirectly to control pollution of coastal waters.

5.2.1 Development in harbours

Development of land in harbour areas must have the permission of harbour authorities and the appropriate planning authority.[21] But much development by harbour authorities themselves in their own areas is exempt from the provisions of the Local Government (Planning and Development) Acts 1963 and 1976.[22] Instead of applying for planning permission, a harbour authority must seek the authorisation of the Minister for Transport for proposed developments under section 138 of the Harbours Act 1946. If the Minister decides to permit the proposed development, he makes a harbour works order. He may also make a harbour works order on his own initiative under section 134 of the Act. A harbour works order may include such 'supplemental and ancillary provisions and such conditions and restrictions as the Minister thinks proper'.[23] In theory, therefore, the Minister has power to attach pollution control conditions to harbour works orders.

The Act provides that public notice be given of the making of a provisional or of a proposed harbour works order;[24] that the order be made available for public inspection; and that written objections and representations with respect thereto be made to the Minister.[25] Whenever the Minister proposes to make a harbour works order, he may, if he thinks fit, direct that a local inquiry be held in regard to the proposed order.[26]

There has been a good deal of public opposition in recent years to the privileged immunity which harbour authorities and the Commissioners for Public Works enjoy from the provisions (especially the citizen-participation provisions) of the Local Government (Planning and Development) Acts 1963 and 1976.[27]

5.3 FISHERIES (CONSOLIDATION) ACT 1959[28]

Section 171 of the Fisheries (Consolidation) Act 1959 applies in respect of discharges to coastal waters; 'waters' for the purpose of this Act includes estuaries and any part of the sea.[29] Although the repeal of this section is provided for in section 34(c) of the Local Government (Water Pollution) Act 1977, it has been continued in force pending the enactment of comprehensive legislation on dumping at sea. Briefly, section 171 prohibits the entry of deleterious matter into any waters except under or in accordance with a licence granted by the Minister for Fisheries. At present four licences to dump waste matter at sea are held under section 171. In only one instance are wastes (approximately 6000 gallons per day of organic sludges arising from the treatment of whey-processing waters) dumped inside territorial waters. The other three licences permit dumping outside the three-mile limit. These licences cover the dumping of waste mycelium (approximately 800,000 gallons per day), dilute spent caustic soda (approximately 1100 cubic metres) and domestic septic tank waste and blood (approximately 250 tons per month).[30]

Section 253 of the Act, as amended,[31] although primarily intended for the protection of molluscs, is in effect a prohibition of certain types of pollution because it makes it an offence to 'deposit ballast, rubbish or other substances' or 'to place any implement, apparatus or thing prejudicial or likely to be prejudicial to any oyster bed or oysters or brood or spawn thereof or oyster fishery, except for a lawful purpose of navigation or anchorage' within the limits of a licensed oyster bed. This section was extended to include mussel, periwinkle and cockle fisheries by section 281 of the Act. Section 290 as amended[32] makes it an offence to discharge ballast from a vessel within any estuary, harbour or place unless such discharge is lawfully permitted. The penalties on conviction for an offence under sections 253 and 290 is a fine not exceeding £500 and £200 respectively.

5.4 LOCAL GOVERNMENT (PLANNING AND DEVELOPMENT) ACTS 1963 AND 1976[33]

Although the functional areas of local planning authorities do not extend beyond the foreshore, since 1976 they are empowered to take into account the 'probable effect' which a particular decision by them on a planning application would have on 'any area which is outside their area'. This presumably means that conditions designed to prevent pollution of coastal waters attached to planning permissions are *prima facie* valid. Planning authorities have been specifically empowered to include objectives to control pollution of the seashore in their development plans.[34]

5.5 LOCAL GOVERNMENT (WATER POLLUTION) ACT 1977[35]

'Waters' are defined in section 1 of the Local Government (Water Pollution) Act 1977 so as to include 'any tidal waters' and, where the context permits, 'any beach, river bank and salt marsh or other area' which is contiguous to tidal waters. Tidal waters are further defined so as to include 'the sea and any estuary up to high water medium tide and any enclosed dock adjoining tidal waters'. Controls under the Act are therefore applicable to both inland and coastal waters, except where otherwise expressly provided. However, where local or sanitary authorities or the Minister for the Environment have power to exercise a discretion it might reasonably be expected that greater latitude will be exercised with respect to discharges of polluting matter to coastal rather than inland waters because of the superior assimilative capacity of the former.

5.6 CONTROL OF POLLUTION FROM OFFSHORE EXPLORATION AND DEVELOPMENT

The rights of the State outside territorial waters over the sea bed and subsoil for the purpose of exploring such sea bed and subsoil and exploiting their natural resources are vested in the Minister for Energy under the Continental Shelf Act 1968.[36] Orders designating where these rights may be exercised are made from time to time under section 2 of this Act.[37] Section 4 extended the application of the Minerals Development Act 1940 and the Petroleum and Other Minerals Development Act 1960 to the continental shelf. These statutes are currently the only legislative measures under which offshore exploration and development may be regulated. Because of their relative antiquity they do not contain many provisions on pollution control. It is unlawful to engage in any activities prospecting for or exploiting minerals or petroleum without an appropriate authorisation under the Minerals Development Act 1940, or the Petroleum and Other Minerals Development Act 1960. When granting any authorisation the Minister for Energy has a discretion to attach conditions, including conditions relating to pollution control. He also has a discretion to revoke any authorisation for breach or non-observance of the terms therein.

Under the Minerals Development Act 1940, every prospecting licence must contain an indemnity clause whereby the licensee undertakes to indemnify the Minister against any claim or demand whatever either in respect of the land or minerals which are the subject of such licence or in any way arising out of the exercise by the licensee of any of the rights conferred on him by such licence.[38] A similar clause must appear in every petroleum exploration and prospecting licence.[39] These indemnity clauses could, in theory at any rate, be drafted so as to enable the Minister to require indemnification against claims for environmental damage caused by the licensee. Mineral and petroleum prospecting licences must also contain clauses requiring the licensee to exercise the rights conferred on him so as not to interfere unnecessarily with the amenities of the area.[40] Compensation must be paid in a manner provided for in the Acts when damage is caused to water supplies or to mineral deposits or whenever a nuisance is caused by a licensee or lessee by the exercise of his rights under a minerals or petroleum prospecting licence or a minerals or petroleum lease.[41] Leases of State minerals or of petroleum must contain such covenants, conditions and subsidiary agreements as the Minister shall consider proper or desirable in the public interest and shall agree with the lessee.[42] These could therefore contain conditions for pollution control although in practice this is not a primary nor indeed a major concern of the Department of Energy. Model terms and conditions applicable to the grant of exclusive licences for exploration for oil and gas on the Continental Shelf were published in a booklet, *Ireland Exclusive Offshore Licensing Terms*. Sections 39, 40, 42, 48, 50, 55, 60 and 65 of the Terms require licensees, *inter alia*,

(i)   to use methods and practices customarily used in good oilfield practice for confining petroleum obtained from the licensed area;

(ii)   to prevent the escape or waste of petroleum or other toxic substances to waters in or in the vicinity of the licensed area which would tend to pollute land or water or damage aquatic life or wild-life or public or private property;

(iii)   to remove pollution caused by drilling or production operations where it damages or threatens to damage aquatic life, wild-life or public or private property;

(iv)   to notify the Minister of any event causing the escape or waste of petroleum.[43]

The Minister has powers:

(i)   to require a licensee to take out appropriate insurance against liability for pollution damage;

(ii)   to require that operations be discontinued or continued subject to conditions where he is satisfied that such a requirement is desirable in order to reduce the risk of damage to the environment;

(iii)   to direct a licensee to remove installations or facilities for environmental or safety reasons when a licence expires or is determined or revoked or when an area is surrendered by the licensee;

(iv)   to require a licensee to indemnify him against any claim, demand or damage arising out of his operations under the licence for injury or damage to persons or property.

(v)   to revoke a licence.

In addition to the above controls, the Department of Energy requires compliance with *Rules and Procedures for Offshore Petroleum Exploration Operations* and *Rules and Procedures for Offshore Petroleum Production Operations*. These require, *inter alia*, that blow-out preventers and auxilliary equipment be installed prior to drilling being carried out and that the well be provided with the necessary casing and that casing strings be run and cemented in such a manner that all uncontrolled movements of fluids into boreholes are avoided. The operator is required to ensure that:

(i)   oil in any form is not disposed of into the sea;

(ii)   liquid waste containing substances which may be harmful to aquatic life or wild-life is treated so as to avoid the disposal of harmful substances into the sea;

(iii)   drilling fluid containing oil is not disposed of into the sea;

(iv)   drilling fluid containing toxic substances is neutralised;

(v)   drill cuttings containing oil are not disposed of into the sea;

127

(vi)   solid waste material is incinerated or brought to shore for disposal.

Offshore operators are also required to prepare plans under the general surveillance of the Department of Energy for dealing with oil spills.[44]

5.6.1 Oil Pollution Liability Agreement (OPOL)

Owners and operators of offshore facilities (including pipelines) in Ireland and several other Western European countries have agreed to operate a strict liability scheme for the payment of pollution damage claims and pollution mitigation expenses. Parties to the OPOL agreement become members of the OPOL Association. The Association must be provided with a certificate of insurance or a guarantee or a surety bond or proof of qualification as a self-insurer indicating an ability to pay $25 million for any one pollution occurrence or $50 million a year. 'Pollution damage' covered is 'direct loss or damage by contamination which results from the discharge of oil . . . from an offshore facility'. OPOL parties agree to reimburse the costs of remedial measures taken by *governmental* authorities and to pay compensation for pollution damage up to $25 million per incident on a strict liability basis. Half of the $25 million is payable for remedial measures and half for pollution damage but, if one type of claim has not exhausted its full $12.5 million, the other type may receive $12.5 million plus the unexhausted amount remaining after the first claim has been satisfied. Amounts spent by an OPOL party in taking remedial action are deductible.

Excluded liabilities are fines, punitive damages, damage to the insured's facilities, the costs of controlling blow-outs, claims directly or indirectly happening through or in consequence of an act of war, insurrection or Act of God, transport of petroleum at sea, and payments made as guarantees by members of the OPOL Association in the event of the default of one of the members. Furthermore, liability is excluded when the polluting incident is caused by an act or omission of a third person with intent to cause damage or wholly caused by the negligence of any government or as a result of compliance with conditions imposed or instructions given by the government of the State which issued the licence to the offshore facility involved. Incidents caused by the intentional acts or negligence of the claimant are excluded (as to that claimant) to the extent that his actions were a cause of the incident.

5.6.2 Penalties for causing pollution

Section 7 of the Continental Shelf Act 1968 provides that if any oil to which section 10 of the Oil Pollution of the Sea Act 1956 applies or any mixture containing not less than one hundred parts of such oil in a million parts of the mixture is discharged or escapes into any part of the sea:

(i)   from a pipeline; or

(ii)   otherwise than from a ship, as a result of any operation for the exploration of the sea bed and subsoil or the exploitation of their natural resources in a designated area,

the owner of the pipeline or, as the case may be, the person carrying on the operations shall be guilty of an offence unless he proves, in the case of a discharge from a place in his occupation, that it was due to the act of a person who was there without his permission (express or implied) or, in the case of an escape, that he took all reasonable care to prevent it and that as soon as practicable after it was discovered all reasonable steps were taken for stopping or reducing it. A person guilty of an offence under section 7 shall be liable on summary conviction to a fine not exceeding £100 and on conviction on indictment to a fine of such amount as the court may consider appropriate. Failure to report to the Minister for Transport discharges of oil from vessels, places or apparatus into the territorial seas is an offence under section 11 of the Oil Pollution of the Sea Act 1977, and is punishable by a maximum fine of £500 and/ or 12 months' imprisonment.

The Petroleum and Other Minerals Development Acts 1940 and 1960, and the Continental Shelf Act 1968, are enforceable by the Minister for Energy. The Oil Pollution of the Sea Acts 1956 to 1977 are enforceable by the Minister for Transport. To date there have been no reported incidents of pollution as a result of offshore exploration and development. Further restrictions on dumping at sea from marine structures have come into force under the Dumping at Sea Act 1981. These are described below.

5.7 DUMPING AT SEA

The legislative controls on dumping at sea contained in the Fisheries (Consolidation) Act 1959, and the Petroleum and Other Minerals Development Acts 1940 and 1960, have already been described.[45] In addition, Dublin Corporation dumps sewage sludge at sea as permitted by the Local Government Board (Ireland) Provisional Order Confirmation (No. 10) Act 1892. Ireland has ratified the Convention on the Prevention of Marine Pollution by Dumping of Wastes and Other Matter 1972 (the London Convention) and the Convention for the Prevention of Marine Pollution from Ships and Aircraft 1973 (the Oslo Convention) by the Dumping at Sea Act 1981.

Under section 3 of the Act the Minister for Transport, after consultation with the Ministers for the Environment, Fisheries, Industry, Commerce and Tourism, and Energy, may grant or refuse to grant a permit in relation to a specified vessel, aircraft or marine structure, authorising the dumping of 'a specified quantity of a specified substance or material in a specified place within a specified period of time, being a substance or material that is intended to be dumped from the vessel, aircraft or marine structure'. In deciding whether or not to grant a permit the Minister must take into consideration the provisions of Annex III to the London Convention. If the permit would relate to a place in the area to which the Oslo Convention applies he must also take into consideration the provisions of Annex III to that Convention. The Minister may attach such conditions as he thinks appropriate to any permit granted and, after consultation with the other Ministers, may revoke or amend a permit whenever he thinks it appropriate to do so. Provision is made for charging fees to cover the costs of tests and investigations to enable the Minister to decide whether to grant or refuse a permit and to allow him, in cases where he proposes to grant a permit, to charge such fees as he thinks appropriate, having regard to the cost of any monitoring, surveys and examinations carried out or to be carried out for the purpose of determining where dumping may take place and to assess the effects of the dumping on the marine environment and the living resources which it supports.

Under section 2 of the Act, it is an offence—punishable on summary conviction by a fine not exceeding £500 and/or 6 months' imprisonment or, on conviction on indictment, to an unlimited fine and/or 5 years' imprisonment—to dump any substance or material:

(i)   in the territorial seas of the State;

(ii)   from an Irish vessel, an Irish aircraft or an Irish marine structure anywhere at sea outside the territorial seas;

(iii)   or to load any substance or material for dumping on a vessel, aircraft or marine structure in the State or Irish territorial waters.

A person charged with the above offence may plead as a defence:

(i)   that the act constituting the offence was carried out by him in pursuance of instructions given to him by his employer;

(ii)   that the commission of the offence was due to a mistake or to the act or default of another person or to an accident or some other cause beyond his control and that he took all reasonable precautions and exercised all due diligence to avoid the commission of such an offence by himself or any person under his control;

(iii)   that the dumping or loading concerned took place under and in accordance with a permit granted by the Minister for Transport or an authorisation granted by another State that is party to the Oslo or London Conventions;

(iv)   that the dumping was reasonably necessary for the purpose of securing the safety of a vessel, aircraft or marine structure or of saving life. In this event, the dumping must be reported to the Minister for Transport as soon as may be but not later than 7 days after it takes place.[46]

The Act will be enforced by officers appointed by the Minister for Transport; by harbour authorities (within their individual jurisdictions); by the Commissioners of Public Works (for harbours of the Commissioners); and Coras Iompair Eireann (for harbours managed by them). It will not apply to State ships or service aircraft. A register of permits granted under the Act must be kept available for public inspection.

The Act is seriously defective in many respects. It allows the Minister to revoke or amend a permit 'whenever he thinks it appropriate to do so' without giving any guidance on circumstances which might justify revocation or amendment of a permit. The Minister is not obliged to give reasons for his decision or to allow disappointed applicants for permits (or holders of permits which have been revoked or amended) to appeal against his decisions. There is no provision for citizen participation on decisions on permits, apart from a requirement that a register of permits issued be kept open for public inspection.

5.8 MISCELLANEOUS

Section 19 of the Public Health (Ireland) Act 1878, which requires sanitary authorities to purify sewage before discharging it, does not apply in respect of discharges to coastal waters. The nuisance provisions of the same Act do not apply to ships and vessels in any river, harbour or district within the jurisdiction of a sanitary authority. Where any river, harbour or water is outside the jurisdiction of a sanitary authority, ships and vessels therein are, unless otherwise provided for by the Minister for the Environment, deemed to be within the jurisdiction of the nearest sanitary authority.[47]

It is an offence under section 4 of the Alkali etc. Works Regulation Act 1906 to deposit or discharge alkali waste without the best practicable means being used for effectually preventing any nuisance arising therefrom. The penalty is £20 in the case of a first offence, £50 for subsequent offences and £5 for every day on which a subsequent offence continues.

Under article 71 of the Fishery Harbour Centres (Management, Control, Operation and Development) By-Laws 1979, made under section 4(2)(*a*) of the Fishery Harbour Centres Act 1968, it is an offence to throw, discharge or suffer any substance other than surface water into the sea within a port without the consent of the Minister for Fisheries. For the purpose of the by-laws 'substance' is defined as including polluting matter, sewage and trade effluent within the meaning of the Local Government (Water Pollution) Act 1977, and bilge water.

5.9 INDIVIDUAL RIGHTS

The principal controls over pollution of coastal waters (other than the Local Government (Planning and Development) Acts 1963 and 1976 and the Local Government (Water Pollution) Act 1977) are administered by central government. The Fisheries (Consolidation) Act 1959, the Petroleum and Other Minerals Development Acts 1940 and 1960, and the Dumping at Sea Act 1981 are administered by the Departments of Fisheries, Energy and Transport respectively.

As is usual in legislation administered by Central Government, the individual has no special rights to enforce the provisions in, or to participate in decision-making on, the various authorisations granted under the Fisheries (Consolidation) Act 1959, the Petroleum and Other Minerals Development Acts 1940 and 1960, and the Dumping at Sea Act 1981.

5.10 COASTAL WATER QUALITY STANDARDS

The only mandatory coastal water quality standards are those contained in the EEC Council Directive 76/160/EEC concerning the quality of bathing water and EEC Council Directive 79/923/EEC on the quality required of shellfish waters.[47] All other standards are determined at the discretion of the various authorities charged with the administration of legislation regulating pollution control of coastal waters.

Among the bodies monitoring the quality of coastal waters are the Eastern and Western Health Boards which monitor compliance with the directive concerning the quality of bathing water; the National Radiation Monitoring Service which monitors radioactivity in the Irish Sea; and the Institute for Industrial Research and Standards which carries out baseline monitoring of coastal and estuarine waters in relation to specific industrial projects and monitoring commissions for public and private sector clients.[48] A number of research organisations have also surveyed pollution levels of coastal waters from time to time but there has been no attempt, as yet, to carry out a comprehensive baseline survey of water quality in Irish coastal waters.[49]

Notes

1. Foreshore Act 1933, s. 5.
2. *Ibid.*, s. 1.
3. Minerals Development Act 1940, s. 82.
4. *Dail Reports*, Vo. 262, Cols 495–497, 4 July 1972.
5. Harbours Act 1976.
6. Foreshore Act 1933, s. 1.
7. *Ibid.*
8. *Ibid.*
9. Harbours Act 1946, ss. 2(3), 47.
10. *Ibid.*, s. 47(1).
11. *Ibid.*, s. 47(2).
12. *Ibid.*, s. 48.
13. *Ibid.*, ss. 50, 52. See also Oil Pollution of the Sea Act 1956, s. 13.
14. *Ibid.*, s. 56.
15. *Ibid.*, s. 58.
16. *Ibid.*, s. 60 and Second Schedule.
17. *Ibid.*, s. 65.
18. *Ibid.*, s. 67.
19. *Ibid.*, ss. 86, 87.
20. *Ibid.*, s. 58.
21. See Chapter 2.
22. See Local Government (Planning and Development) Regulations 1977, Third Schedule, Class 20.
23. Harbours Act 1946, s. 134(5).
24. *Ibid.*, ss. 135, 136.
25. *Ibid.*
26. *Ibid.*, s. 137.
27. See, for example, *Hibernia*, 12 July 1979.
28. See 4.2.
29. Fisheries (Consolidation) Act 1959, s. 3.
30. *Dail Reports*, Vol. 319, Cols 1343, 18 April 1980.

31. Fisheries (Amendment) Act 1962, s. 32. Fisheries Act 1980, s. 50.
32. *Ibid.*
33. See Chapter 2.
34. Local Government (Planning and Development) Act 1963, Third Schedule, Part IV, 11.
35. See 4.4.
36. Continental Shelf Act 1968, s. 2.
37. Continental Shelf (Designated Areas) Orders 1968, 1970, 1974, 1979.
38. Minerals Development Act 1940, s. 8(4).
39. Petroleum and Other Minerals Development Act 1960, ss. 8(5), 9(6).
40. Minerals Development Act 1940, s. 8(5); Petroleum and Other Minerals Development Act 1960, s. 9(7).
41. Minerals Development Act 1940, s. 10, 25, 31 and Petroleum and Other Minerals Development Act 1960, ss. 12, 16, 28.
42. Minerals Development Act 1940, s. 26; Petroleum and Other Minerals Development Act 1960, s. 13(2).
43. See *Ireland Exclusive Offshore Licensing Terms*, Prl. 4510, Sections 39, 40, 42, 48, 50, 55, 60, 66.
44. Information from Department of Energy.
45. See 4.2 and 5.6.
46. Dumping at Sea Act 1981, s. 3.
47. See 4.6.
48. *Report on Pollution Control*, pp. 52–56.
49. See Toner and O'Sullivan, *Water Pollution in Ireland* (1977), National Board for Science and Technology, Dublin, pp. 45, 46, 48, 49.

6

Oil Pollution of the Seas

In the last six years there have been three major and several minor oil spills in Irish waters. The three major spills all occurred in Bantry Bay, the major oil trans-shipment terminal built in 1968 to accommodate large oil tankers. Tanker movements in Ireland are of three kinds: (i) tankers loading and unloading at Bantry, (ii) transit traffic passing through Irish territorial waters, and (iii) the movements around the coast of domestic tankers which are sent from the refinery at Whitegate to Drogheda, Dublin, Arklow, Tarbert, Foynes and Galway.

Since oil pollution often occurs beyond the limits of national jurisdiction and can involve ships from many States, it is controlled largely by International Conventions implemented on a national level by domestic legislation.

Responsibility for controlling or dealing with oil pollution is divided between several Ministers. The Minister for Transport has overall responsibility for implementing the Harbours Acts 1946–79 and the Oil Pollution of the Seas Acts 1956–77. The Minister for Energy is responsible for the prevention of oil pollution from offshore exploration and development under the Continental Shelf Act 1968. The Minister for Labour has a general responsibility for implementing the Dangerous Substances Act 1972, which, *inter alia*, operates to prevent oil pollution from activities at oil jetties and petroleum stores and from dealing with petroleum. The Minister for Fisheries has a general responsibility for safeguarding fishing interests under the Fisheries Acts 1959–80.

Overall responsibility for arrangements for cleaning up oil pollution rests with the Department of the Environment but local authorities clean up oil from beaches and immediately offshore and harbour authorities deal with oil spillages in harbours. The Department of Defence is responsible for cleaning up oil at sea.

6.1 CRIMINAL LIABILITY FOR OIL POLLUTION

The most important legislative provisions regulating criminal liability for oil pollution are sections 10 and 11 of the Oil Pollution of the Sea Act 1956, as amended by the Oil Pollution of the Sea Acts 1965 and 1977.

6.1.1 Discharges from Irish registered ships

Section 10(3) of the 1956 Act (as amended[1]) provides that:

> The owner and also the master of any ship registered in the State which discharges oil or any mixture containing oil, anywhere at sea shall be guilty of an offence.

It should be noted that this subsection explicitly imposes liability on both the owner and the master of the ship. This formulation avoids the interpretational difficulties which arose in England in *Federal Steam Navigation Co. Ltd.* v. *Department of Trade and Industry*[2] where the wording of a section similar to section 10(3) imposed liability on the owner *or* the master. The Oil Pollution of the Sea Acts do not contain any definition of a ship, but section 742 of the Merchant Shipping Act 1894 provides that 'a ship includes every description of vessel used in navigation not propelled by oars'. A 'discharge' is defined as 'any discharge or escape howsoever caused'.[3] Section 10 applies to crude oil, fuel oil, lubricating oil, heavy diesel oil or 'other description of oil' which may be prescribed by Ministerial regulation. The Minister for Transport has extended the applications of s. 10(3) to diesel oil other than distillates of which more than 50% by volume distills at a temperature not exceeding 340 °C when tested by the American Society for Testing Materials Standard Method D.86/59", by article 2 of the Oil Pollution of the Sea Act 1956 (Application of Section 10) Regulations 1980.[4] 'Oily mixture' is not defined in current legislation on oil pollution but presumably the commonsense definition contained in the 1969 Amendment to the 1954 Convention (i.e. a mixture with any oil content) would be accepted by an Irish Court.

Section 10 is confined in its application to Irish registered ships because of the current interpretation of rules of international law on flag State jurisdiction which prohibit a State, other than that where the ship is registered, from prosecuting for an offence committed outside its jurisdiction. However, an *Irish* registered ship which commits the offence in section 10 can be prosecuted for so doing irrespective of whether the offence is committed on high seas or in the territorial seas of another State. In practice, however, it is likely that in the latter case enforcement of oil pollution legislation would be left to the injured State.

Section 10(5) provides that the Minister for Transport may prescribe exceptions to section 10(3)

> either absolutely or on specified conditions and either generally or for specified classes of ships, or in relation to particular descriptions of oil or oily mixtures, or to their discharge in specified circumstances, or in relation to particular sea areas.

The Minister has made regulations under section 10(5) of the Oil Pollution of the Sea Act 1965 (Exceptions) Regulations 1980,[5] article 3 of which provides that ships other than tankers and tankers in relation only to discharge of oil or mixtures containing oil from their machinery space bilges shall be exempted from the operation of section 10(3) of the Oil Pollution of the Sea Act 1956, as amended,[6] provided certain specified conditions are complied with. Different conditions apply depending upon whether the vessel concerned is a ship or a tanker as defined in article 2 of the regulations.[7]

6.1.2 Discharges of oil into Irish national waters

Section 11 of the Oil Pollution of the Sea Act 1956, as amended,[8] provides that:

> (1)(a) If any oil or oily mixture is discharged (directly or indirectly) into the territorial seas of the State, or into any of its inland waters that are navigable by sea-going vessels, or on its seashore, then, if the discharge is:
>
> (i)   from a vessel, the owner and also the master of his vessel,
>
> (ii)   from a place at land, the occupier of that place,
>
> (iii) from an apparatus for transferring oil to or from a vessel, the person in charge of the apparatus, shall be guilty of an offence.
>
> (b)   In this subsection 'seashore' has the same meaning as in the Foreshore Act 1933.

Unlike section 10(3) which is specifically limited in its application to certain types of oil, section 11 relates to oil and oily mixtures as defined in section 3 of the Oil Pollution of the Sea Act 1956, i.e. 'oil of any description, and includes spirit produced from oil, and coal tar'. This is a somewhat wider definition of oil than that adopted in the 1954 Convention as amended in 1969 and it is unlikey that the problems which arose in *Cosh* v. *Larsen*[9] would ever arise in Ireland.

The offence under section 11 consists of polluting Irish waters and— since 1977—the seashore, by oil or oily mixture. The nationality of the polluting ship is irrelevant because in this case (unlike that covered by section 10) the State is not inhibited by rules of international law from prosecuting in respect of offences committed within its own jurisdiction.

The provision that the State may prosecute in respect of oil pollution of the seashore—an innovation introduced by the 1977 Act—although not paralleled in the Convention for the Prevention of the Pollution of the Sea by Oil 1954, as amended, was necessary to provide a sound statutory basis for prosecuting in respect of pollution of the seashore. Since the penalties for oil pollution have been increased, it is possible that a penalty imposed might be sufficient to compensate for damage to the seashore.

The discharge of oil or oily mixture to inland waters navigable by sea-going vessels is also an offence. The question of whether a particular area comes within the definition of 'inland waters navigable by sea-going vessels' has caused difficulty on at least one occasion, in 1974 in the port of Dublin, but the recent decision in *Ranklin* v. *Da Costa*[10] should resolve some of the difficulties surrounding this question in that it has established that the test of navigability should be primarily a geographical rather than a functional one.

6.1.3 Exceptions to sections 10 and 11 of the 1956 Act

Section 12(1) of the 1956 Act provides that sections 10 and 11 thereof shall not apply to:

(*a*) the discharge of oil or an oily mixture from a vessel for the purpose of securing the safety of the vessel, preventing damage to the vessel or her cargo, or saving life, if such discharge was necessary and reasonable in the circumstances, or

(*b*) the escape of oil or of an oily mixture from a vessel, resulting from damage to the vessel or from any leakage, not due to want of reasonable care, if all reasonable precautions have been taken after the occurrence of the damage or discovery of the leakage for the purpose of preventing or minimising the escape.

Section 12(2) provides that section 11 is not to apply to the discharge from any place of an effluent produced by operations for the refining of oil, if:

(*a*) it was not reasonably practicable to dispose of the effluent otherwise than by so discharging it, and

(*b*) all reasonably practicable steps had been taken for eliminating oil from the effluent, and

(*c*) in the event of the surface of the waters into which the mixture was discharged, or the land adjacent to those waters being fouled by oil at the time of discharge, it is shown that the fouling was not caused or contributed to by oil contained in any effluent discharged at or before that time from that place.

There is only one oil refinery in Ireland—at Whitegate, County Cork. About 40–60% of the nation's oil is refined there.

6.1.4 Penalties for oil pollution

Section 23 of the 1956 Act, as amended,[11] provides that the penalty on conviction for an offence under section 10 or 11 is:

(i)   on summary conviction, a fine not exceeding £500, or imprisonment for a term not exceeding 12 months, or both such fine and such imprisonment;

(ii)   on conviction on indictment, a fine not exceeding £100,000 together with £10,000 per day for a continuing offence.

There is a limit to the maximum fine which can be imposed for offences under sections 10(3) and 11(1). Considering the damage which oil pollution can cause, it is questionable whether these fines are, in the words of article IV of the 1962 Amendments to the 1954 Convention, 'adequate in severity to discourage any such unlawful discharge'. Certainly the owner and/or the master of a ship who wished to effect an unlawful discharge of oil would be well advised, if he had a choice, to do so in Irish rather than in UK waters as by doing so he would reduce his liability for heavy fines and, indeed (in view of the level of enforcement of oil pollution controls in Ireland), he would almost certainly escape being prosecuted at all.

6.1.5 Miscellaneous

Criminal prosecutions for oil pollution of harbour waters may also be brought by harbour authorities under section 88 of the Harbour Acts 1946, and by the Minister for Transport in respect of oil pollution of the foreshore under the Foreshore Act 1933. These Acts are never used to enforce oil pollution controls. Sections 3 and 4 of the Local Government (Water Pollution) Act 1977 do not apply in respect of discharges to tidal waters from vessels or marine structures.[12]

6.2 CIVIL LIABILITY FOR OIL POLLUTION

6.2.1 Legislation on civil liability for oil pollution

No legislation has been enacted in Ireland for the specific purpose of regulating civil liability for oil pollution damage. Ireland has not yet ratified the International Convention on Civil Liability for Oil Pollution Damage 1969, nor the International Convention on the Establishment of a Fund for Compensation for Oil Pollution Damage 1971. The only legislative provision relative to the question of civil liability for oil pollution is section 13 of the Local Government (Water Pollution) Act 1977, which provides that:

(1) Where it appears to a local authority that urgent measures are necessary to prevent pollution of any waters in its functional area, to remove polluting matters from waters in that area, or where such matter is in waters outside that area, to prevent it affecting any part of that area, the local authority may take such steps, carry out such operations or give such assistance as it considers necessary to prevent such matters from entering the waters, to remove the matter from the waters, to dispose of it as it thinks fit and to mitigate or remedy the effects of any pollution caused by the matter.

(2) Where a local authority takes steps, carries out operations or gives assistance under this section it may recover the cost of such steps, operations or assistance as a simple contract debt in a court of competent jurisdiction from such person as the local authority

satisfies the court is the person whose act or omission necessitated such steps, operations or assistance.

Under the arrangements made by the Department of the Environment for the clearance of oil pollution from coastal areas, maritime local authorities are responsible for clearance of oil from beaches and immediately offshore.[13] Before the enactment of section 13, the costs of clean-ups would be recoverable (if at all) by local authorities suing for negligence, nuisance, trespass.[14] But since 1977 local authorities may recover their costs from the persons responsible provided that (a) they considered that urgent measures were necessary to prevent oil pollution, and (b) waters in their functional area were threatened or polluted by oil. It should be noted that section 13 allows local authorities to claim the costs of preventive as well as curative actions and that the powers given extend to operations outside the functional area of the local authority (e.g. the sea) where waters in a local authority's own area are at risk. A local authority has never yet brought proceedings to recover costs under section 13. It is unfortunate that section 13 was not framed so as to cover steps taken by harbour authorities and other bodies charged with preventing or cleaning up oil pollution.[15]

6.2.2 The common law on civil liability for oil pollution[16]

Apart from section 13 the law on civil liability for oil pollution is the general law of tort. This, however, is peculiarly unsuitable as an instrument for recovering damages for oil pollution. Actions may be brought for trespass but there are difficulties in that the trespass must be direct and intentional and it appears that the courts, since *Esso Petroleum Co. Ltd.* v. *Southport Corporation*,[17] consider that interference by oil pollution from a ship at sea is not direct and therefore trespass does not lie. *Rylands* v. *Fletcher* may not be used to ground an action for oil pollution originating from the sea as this tort only applies in respect of escapes from land. A person seeking damages for negligence could run into problems of proving causation and establishing reasonable foreseeability. Actions for private nuisance may only be brought by plaintiffs who have suffered injury to their proprietary interests in land, while most oil pollution affects the foreshore (most of which is owned by the State) or the seashore where damage caused is not easily quantifiable or of a commercial nature. The Attorney General could possibly bring an action for public nuisance but it would be more difficult for the private citizen to do so unless he could prove that he had suffered special damage. The possibilities of successfully suing at common law for damage caused by oil pollution are therefore quite limited.

6.2.3 Industry practice: TOVALOP and CRISTAL

TOVALOP and CRISTAL are private insurance schemes voluntarily established by those involved in the oil industry to settle claims in respect of civil liability for oil pollution damage without resort to litigation.

TOVALOP (Tanker Owners Voluntary Agreement Concerning Liability for Oil Pollution) provides a private fund for compensating oil pollution damage for which *tanker owners* are liable.[18] It is administered by the International Tanker Owners Pollution Federation Ltd. in London. Over 90% of tanker owners are parties to the agreement. Parties bind themselves for short periods to pay claims for oil pollution damage up to a certain limit. Damages are recoverable on a strict liability basis and cover clean-up costs and 'threat removal measures', i.e. 'reasonable measures taken by any person after an incident has occurred for the purpose of removing a threat of an escape or discharge of oil'.[19] Damages are excluded for any loss or damage which is remote or speculative, or which does not result directly from an escape or discharge of oil.[20]

CRISTAL (Contract Regarding an Interim Supplement to Tanker Liability for Oil Pollution) provides a fund for compensating oil pollution damage for which *oil companies* are liable. It is administered by the Oil Companies Institute for Maritime Pollution Compensation Ltd. in Bermuda.

There are numerous exceptions and qualifications to liability under TOVALOP and CRISTAL, but in general they have filled legal lacunae far quicker and perhaps far better than the Irish legislature. All claims by Irish public authorities for compensation for oil pollution damage to date have been paid under TOVALOP.

6.2.4 Experiences in recovering compensation for oil pollution damage

Irish public authorities take the view that primary responsibility for cleaning up oil pollution lies with those responsible for causing it. Both Irish Refining Ltd., which runs the oil refinery at Whitegate, and Gulf Oil Ltd., which operates the oil terminal at Bantry Bay, have personnel, facilities and equipment for cleaning up oil spills in Cork harbour and Bantry Bay respectively. There have been numerous minor and three major oil spills off the Irish coast but no compensation for damage caused has ever been recovered in civil proceedings. A number of claims for compensation have, however, been settled out of court. The clean-up costs after the large *Universe Leader* and *Afran Zodiac* spills in 1974 and 1975 were paid by Gulf Oil. Cork County Council (the public authority most directly affected) was also paid £17,800 for out-of-pocket expenses. The bill for expenses incurred by Irish public authorities co-operating with British authorities dealing with the *Christos Bitas* incident[21] in October 1978 was submitted by the UK Department of Trade on Ireland's behalf to TOVALOP (Ireland could not submit the bill herself because she had not ratified the Convention on Civil Liability for Oil Pollution Damage 1969).

The Department of Transport has assumed responsibility for coordinating all claims by public authorities for losses occasioned as a result of oil pollution, and for presenting the bill to the polluter and/or his insurers.[22] The Department has not issued any guidelines on what losses are or are not recoverable and no policy on this important issue has been considered or formulated.

Where civil liability for oil pollution damage cannot be established (e.g. where the polluter is untraceable or has a good defence to his action) clean-up costs must be borne by the public authorities. The Department of the Environment has agreed that the State will meet 50% of the cost to local authorities of clearance of oil pollution from beaches and immediately offshore, including the cost of the purchase of approved dispersant, equipment and other materials, subject to a limit on local authority expenditure for this purpose of a sum equal to the produce of 2½ pence in the pound in the rates of any year: expenditure over this limit in any year will be fully subsidised.[23] Expenses incurred by harbour authorities, other than those recoverable from the polluter, are paid by the State.

6.3 CONTROL OF NAVIGATION

One of the recognised measures for preventing oil pollution is to prevent tankers colliding. Ireland ratified the convention on the International Regulations for Preventing Collisions at Sea 1972, on 12 July 1977, with the making of the Collision Regulations (Ships and Watercraft on the Water) Order 1977[24] made under sections 418 and 424 of the Merchant Shipping Act 1894, as amended. These regulations, *inter alia*, give effect to IMCO traffic separation schemes. Two of these schemes operate in areas around the Irish coast—one at Carnsore Point off Tuscar Rock and the other off Fastnet. Failure to comply with the collision regulations is an offence for which the owner or the master of an offending vessel may be prosecuted.[25]

6.4 CREW STANDARDS

The Merchant Shipping (Certification of Seamen) Act 1979 was enacted *inter alia* to enable Ireland to ratify the International Convention on the Training, Certification and Watchkeeping of Seafarers. The joint Oireachtas Committee on Secondary Legislation which monitors the implementation of EEC law was advised in December 1977 that 'Irish standards are already as high as those provided in the Convention'.[26]

6.5 CONSTRUCTION, EQUIPMENT AND OPERATION OF SHIPS

6.5.1 Implementation of international requirements

The Thirty-Eighth Report of the Joint Committee on Secondary Legislation reported in December 1977 that Ireland was to accede to the Administrative Agreement of 2 March 1978 between eight North Sea countries on the Maintenance of Standards on Merchant Ships and that 'this necessarily involves ratification of ILO Convention No. 147'. Ireland has not yet acceded to this Agreement.

Ireland ratified the International Convention for the Safety of Life at Sea 1960 by the Merchant Shipping Act 1966 and by regulations made thereunder which require that ships be constructed and equipped to certain standards.[27] The International Convention for the Safety of Life at Sea 1974 (SOLAS 1974) and Protocol of 1978 were implemented by the Merchant Shipping Act 1981. The International Convention on Load Lines 1966 was implemented by the Merchant Shipping (Load Lines) Act 1968, and by legislation made thereunder. The Joint Committee on Secondary Legislation was advised that the procedures set out in IMCO Resolution A 321 (IX) are observed by marine surveyors during normal dock inspections. The International Convention for the Prevention of Pollution from Ships 1973 (MARPOL 1973), and Protocol 1978, have not been ratified.

EEC Council Directive 79/116/EEC as amended by 79/1034/EEC, concerning the minimum requirements for certain tankers entering or leaving Community ports, was implemented by the European Communities (Entry Requirements for Tankers) Regulations 1981 which prescribe minimum reporting requirements for certain oil, gas and chemical tankers entering or leaving Community ports.[28] These regulations also implement requirements of the IMCO codes for the Construction and Equipment of Ships Carrying Dangerous Chemicals in Bulk (1977), as amended; for Existing Ships Carrying Liquefied Gases (1976), as amended; and for the Construction and Equipment of Ships Carrying Liquefied Gases in Bulk (1976), as amended. The maximum penalty for breach of the regulations is £500.

It is understood that the Department of Transport proposes to implement the EEC Directive concerning the pilotage of vessels by deep-sea pilots in the North Sea and English Channel by issuing a marine notice for the attention of Irish registered vessels.

A Merchant Shipping Act 1981 was passed in May 1981 to give effect to the International Convention for the Safety of Life at Sea 1974, and to implement recommendations on ship safety and operational procedures recommended in the report of the Tribunal of Inquiry into the Whiddy Island disaster. Section 4 of the Act makes it an offence punishable by a maximum fine of £50,000 for a ship, having regard to the nature of the service for which the ship is intended, to be unfit by reason of the condition of the ship's hull, equipment or machinery or by reason of undermanning or overloading or improper loading to go to sea with serious danger to human life. Both the master and the owner of the ship may be prosecuted. This section applies in respect of Irish registered ships and other ships in Irish territorial waters. Section 5 of the Act empowers the Minister for Transport to make regulations requiring ships to carry such adequate and up-to-date charts, nautical directions and information and other nautical publications as appear to him to be necessary or expedient for the safe operation of ships. In this case the Master *or* the owner may be liable on summary conviction for breach of the regulations to a maximum fine of £500.

6.5.2 Equipment of ships to prevent oil pollution

Section 16 of the 1956 Act, as amended,[29] provides that the Minister for Transport may make regulations

> requiring ships or certain categories of ships registered in the State to be so fitted or constructed, and to comply with such requirements, as to prevent or reduce the discharge of oil or oily mixtures into the sea.

The Oil Pollution of the Sea Act (Ships Equipment) Regulations 1980[30] made under section 16(1) of the 1956 Act as amended provide that every ship registered in Ireland which uses oil as fuel for either engines or boilers must be fitted

> so as effectively to prevent oil fuel from leaking or draining into bilges unless the contents of the bilges are subjected to an effective means of separating the oil therefrom before they are pumped into the sea.

These regulations implement Article VII of the Convention for the Prevention of Pollution of the Sea by Oil 1954, as replaced by Article VII of the 1969 Amendments. Section 16(2) of the 1956 Act provides that a surveyor of ships and such other persons as the Minister may appoint may carry out such tests as may be prescribed for the purpose of section 16. Section 16(3) provides that breach of regulations made under section 16(1) shall be an offence for which both the owner and the master of the ship may be prosecuted. Tests have not been prescribed for the purpose of section 16 and there have never been any prosecutions for breach of the Oil Pollution of the Sea Act (Ships Equipment) Regulations 1957.

Section 18 of the 1956 Act and the 1981 Act and orders made thereunder provide that the Government may by order apply section 16 to ships registered in other countries while they are in a harbour in the State or within territorial seas while on their way to or from such a harbour unless their presence there is due to stress of weather or some other unforeseen and unavoidable circumstance. The Government is also empowered by section 18(2) to exempt foreign ships from the requirements of section 16 if satisfied that the law of the country where a ship is registered is as effective as Irish law and that the ships comply with it. The Government has never made any orders applying section 16 to ships other than those registered in Ireland.

Other requirements are contained in the Merchant Shipping Act 1981, and in the European Communities (Entry Requirements for Tankers) Regulations 1981, referred to in the preceding section.

6.6 FACILITIES FOR THE DISPOSAL OF OILY RESIDUES

Section 13(1) of the 1956 Act empowers (but does not oblige) harbour authorities to provide facilities for the disposal of oily residues from vessels using their harbour. Section 13(2) empowers the Minister for Transport, after consultation with the harbour authority and with any organisation representative of owners of Irish registered ships, to require a harbour authority to provide oil reception facilities or adequate oil reception facilities within a specified time. Under section 5 of the 1956 Act, the Minister for Transport is empowered to require the provision of oil reception facilities or adequate oil reception facilities by oil refineries, ship building and ship repairing yards. Failure to comply with the Minister's requirements is an offence.

The provision of facilities for the disposal of oily residues is one of the most important steps that can be taken in the fight against oil pollution of the sea. Article 8 of the International Convention for the Prevention of Pollution of the Sea by Oil 1954, as amended in 1962, requires the provision of such facilities. The International Convention for the Prevention of Pollution from Ships 1973 requires that the Government of each contracting party undertake to ensure the provision of reception facilities at oil loading ports, terminals, repair ports and in other ports in which the ships have oily residues to discharge.

No facilities for the reception or disposal of oily residues from ships or tankers have been provided by harbour authorities in Ireland. Even the port of Dublin has no such facilities. The Minister for Transport has never exercised his powers under section 13(2) of the 1956 Act to require the provision of oil disposal facilities (it should be noted that, unlike his UK counterpart, the Minister has no power to prosecute harbour authorities for failure to comply with his requirements). However, fixed facilities for the disposal of oily residues have been provided at Whitegate in Cork Harbour and at the Gulf Oil Terminals in Bantry Bay—the country's only oil refinery and the terminal which handles the greatest amount of oil respectively. Mobile facilities are available at the Verholme Dockyard in Cork where ships are built and repaired. All of these facilities are provided by private interests and do not involve any charges on the Exchequer. There are a number of private waste disposal contractors—one of whom specialises in waste oils—who could, if necessary, be employed on a private contractual basis to dispose of oily residues. The overall position appears to be that reception facilities are available at the three places most likely to require them on a regular basis, but not elsewhere. One explanation given for this unsatisfactory state of affairs was the fact that most ships visiting Irish ports do so only to discharge cargoes and so do not need oil disposal facilities. A side effect of the EEC requirements on minimum stocks of oil is that oil companies who import oil to Dublin and Cork have used their entire oil storage capacities to meet EEC obligations and there is no spare capacity which could be used as slop tanks for waste oils.

6.7 OIL RECORDS

Section 17 of the 1956 Act, as amended,[31] and regulations made thereunder implement Article IX of the International Convention for the Prevention of Pollution of the Sea by Oil 1954, as amended in 1962 and 1969. The purpose of section 17 and of the Oil Pollution of the Sea (Records) Regulations 1980[32] is to enable competent authorities in Ireland to tell by inspecting a ship's oil record book whether or not an unlawful discharge of oil has been made. In practice, it appears that little effort is made in Ireland to enforce oil records regulations because of a shortage of technical personnel in the Department of Transport. There has never been a prosecution under section 17.

Section 18 of the 1956 Act empowers the Government to extend the application of section 17(1) to foreign registered ships while they are in a harbour in the State or within the territorial seas while on their way to or from such a harbour, unless their presence there is due to stress of weather or some unforeseen and unavoidable circumstances. No orders have been made under section 18.

Section 19 of the 1956 Act, as amended,[33] makes provision for the enforcement of the 1954 Convention by empowering the Minister for Transport to make orders declaring that a particular country has accepted or denounced the Convention or that the Convention extends or has ceased to extend to any territory. The Minister has made several such orders. Section 19 also provides that a surveyor of ships or any person empowered by warrant of the Minister may go on board any ship to which the Convention applies while she is within a harbour in the State, may inspect and take a copy of any entry in the oil record book, and may require the master of the ship to certify it as a true copy. Any person who impedes a surveyor of ships in the exercise of his functions or who fails to accede to a request to produce an oil record book for inspection or to certify a copy of an entry shall be guilty of an offence and shall, on summary conviction, be liable to a fine not exceeding £500 and/or 12 months' imprisonment. No prosecutions have been brought for breach of section 19.

6.8 CONTROLS OVER SHORE INSTALLATIONS AND PORT ACTIVITIES

A number of Acts and statutory instruments contain controls which operate to prevent oil pollution from shore installations and port activities. The most important of these are as follows.

148

I Harbours Act 1946. Controls over oil pollution of harbour areas in the Harbours Act 1946 are described above.[34]

II Oil Pollution of the Sea Acts 1956 to 1977. The Oil Pollution of the Sea Acts 1956 to 1977 make special provision for the exercise of control by harbour authorities over oil pollution of their areas. Prosecutions for offences created by the Act which occur in harbours may be brought by harbour authorities.[35] A harbour master may specify where, at what times, and under what conditions, ballast water from a petroleum carrying ship may be discharged in harbour waters.[36] All discharges and escapes of oil or oily mixtures from vessels or land into harbour waters must be reported to the harbour master, as must the cause of such occurrences: failure to make a report is punishable by a maximum fine of £100.[37] Oil may not be transferred to or from a vessel in a harbour at night unless notice of the transfer is given to the harbour master. Failure to give such notice is punishable by a maximum fine of £50.[38]

III Dangerous Substances Act 1972. The Dangerous Substances Act 1972 requires that petroleum spirit (as defined in the Act)[39] be kept in stores licensed by the appropriate local or harbour authority.[40] If the store is owned by such authority, the licence must be granted by the Minister for Labour.[41] A local or harbour authority has a discretion to grant or refuse an application for a licence and to attach such conditions as it thinks proper to any licence granted.[42] When attaching conditions, the local or harbour authority is bound to comply with any regulations made by the Minister as to the conditions under which licences may be granted.[43] The Minister also has a discretion to grant or refuse a licence application and to attach such conditions as he thinks proper to any licence.[44] If a local or harbour authority or the Minister think it proper they may publish particulars of a licence application and invite representations concerning it from interested persons.[45] If a local or harbour authority refuses a licence or attaches conditions which are unacceptable to the applicant, he may appeal to the Minister.[46] The appeal must be brought within 10 days after receipt by the applicant of a certificate stating the grounds on which the licence was refused or the conditions attached, unless the Minister allows an appeal outside this time limit.[47] The Minister, having considered the appeal, may, at his discretion, direct the authority to grant or cancel the licence, to attach specified conditions to the licence or to amend or delete a condition attached, and the authority must (unless a successful appeal is taken against the Minister's decision) comply with the Minister's direction.[48] Appeals against the Minister's decision on a licence application or against his direction on appeal from a decision of local or harbour authority may be brought to the High Court.[49] The High Court may direct the Minister to grant the licence, to attach specified conditions or to amend or delete an existing condition, as appropriate.[50] On an appeal from a direction, the Court may either confirm the direction or direct the Minister to vary it.[51]

The Dangerous Substances Act was not brought into force until 18 September 1979. On that day also the Minister for Labour made two sets of regulations, the Dangerous Substances (Retail and Private Petroleum Stores) Regulations 1979[52] and the Dangerous Substances (Petroleum Bulk Stores) Regulations 1979,[53] which, *inter alia*, regulate in great detail the design, construction, maintenance and operation of petroleum stores and ancillary plant and prescribe operational procedures and practices to be observed by those dealing with petroleum.

The Minister has also made the Dangerous Substances (Oil Jetties) Regulations 1979,[54] which provide for all practical steps to be taken to prevent risk of injury to persons or property:

(i)   in the vicinity of petroleum ships and oil jetties;

(ii) in loading or unloading petroleum ships at oil jetties or in harbours;

(iii) in the conveying of petroleum by pipeline to or from ships' tanks and storage tanks ashore.

Operational procedures and practices to be observed in loading or unloading petroleum to ships are prescribed to ensure that oil does not escape. The use of an oil jetty for the purpose of loading and unloading petroleum must be with the consent of the Minister or the appropriate harbour master. Loading and unloading operations carried on at the oil jetty and the conveying of petroleum by pipeline may only be carried out in accordance with the relevant requirements in the regulations.

The conveyance of petroleum by road is also regulated in great detail by the Dangerous Substances (Conveyance of Petroleum by Road) Regulations 1979.[55]

Ships carrying petroleum spirit or other dangerous substances must, on entering any harbour, notify the harbour master of the nature of their cargo. Failure to give such notice is an offence for which the owner and master are liable. It is a good defence to prove that they did not know or could not with reasonable diligence have found out the nature of their cargo.[56]

Contravention of any provision of the Act, or of any instrument thereunder, or of any condition attached to a licence is an offence for which the penalty on summary conviction is a fine of up to £100 and/or 6 months' imprisonment plus £20 for each day a contravention is continued after conviction.[57] Prosecutions may be brought by the Minister or a local or harbour authority, as appropriate.[58] Regulations may provide for the revocation of licences.[59]

6.9 INTERVENTION UNDER THE INTERNATIONAL CONVENTION RELATING TO INTERVENTION ON THE HIGH SEAS IN CASES OF OIL POLLUTION CASUALTIES 1969

This Convention was implemented by the Oil Pollution of the Sea (Amendment) Act 1977 which, together with the Application of Section 2 Order 1979,[60] confers new and extensive powers on the Minister for Transport to deal with maritime casualties likely to result in oil pollution even where such casualties occur outside territorial waters. Action has never yet been taken under the Act.

No administrative arrangements have been made to ensure that the Minister is in a position to exercise his powers under the 1977 Act with the aid of competent advice or advisors. There are no salvage firms in Ireland with whom possible contingency plans could be negotiated nor have any arrangements yet been made with foreign salvage firms.

Although Article 19 of the 1969 Convention requires the coastal State to consult with other States affected and to notify interested parties of the proposed measures which are to be taken except in cases of extreme urgency necessitating immediate action, the Irish Act is, like the equivalent UK legislation, silent on any such duties. No formal arrangements have been made by the Irish authorities as to the manner in which such consultations are to be held or as to how notifications are to be given.

The 1973 Protocol to the Convention extending intervention powers to pollution by substances other than oil has not been ratified by Ireland.

6.10 ENFORCEMENT

It is extremely difficult to get information on the number and effects of oil pollution incidents in Ireland. There are no effective monitoring units for oil pollution and most authorities responsible for enforcing oil pollution legislation treat any enquiries as to how they are exercising their functions with a good deal of suspicion. The Department of Transport is the prosecuting authority in respect of pollution at sea or on the seashore (other than in harbour areas) and harbour authorities are empowered to prosecute in respect of offences committed within harbour jurisdictions.

In the years 1970–73, Cork Harbour Commissioners brought seven, four and one prosecutions respectively under the Oil Pollution of the Sea Act 1956, although in fact there were twenty-four spills in Cork Harbour in that period, eleven of which caused residual pollution. Eighteen spills originated from vessels, three from premises and the sources of three spills were unknown. In the same period Dublin Port and Docks Board brought two prosecutions and the Minister for Transport six. In the period 1975–78 Cork Harbour Commissioners brought three prosecutions. The Department of Transport has not brought any prosecutions under the Acts since 1975. One reason for this is said to be a jurisdictional defect in the 1956 Act but this was remedied in the 1977 Act. The Acts are enforceable on behalf of the Minister of Transport by the Marine Service Section of his Department. The staff in this section consists mostly of surveyors. In 1977, 1978 and 1979 there were six, four and six surveyors respectively employed in the section.[61]

6.11 ADMINISTRATIVE ARRANGEMENTS FOR DEALING WITH OIL POLLUTION

6.11.1 Administrative structure

Irish authorities take the view that the polluter is primarily responsible for cleaning up oil pollution. Accordingly, most of the spills which have occurred in Bantry Bay (including two of the three major spills) have been cleaned up by Gulf Oil Ltd. Similarly, oil pollution at or in the vicinity of the oil refinery at Whitegate is dealt with by Irish Refining Ltd., which operates the refinery. However, where or when the polluter does not, or cannot, clean up oil pollution, this must be done by one or more of the various public authorities with responsibilities for the clearance of oil pollution. Overall responsibility for arrangements for cleaning up oil pollution lies with the Department of the Environment which is also responsible for coordinating the activities of the many other bodies who may become involved in clean-ups. These are:

Local Authorities: clearance of oil from beaches and immediately off-shore.

Harbour Authorities: clearance of oil spillages in harbours.

Department of Defence: clearance of oil at sea.

Department of Energy: control of oil pollution arising from off-shore exploration.

Department of Transport: control of discharges of oil from vessels.

Department of Fisheries and Forestry: protection of fisheries and wild-life.

The activities of all authorities concerned are coordinated by a Liaison Committee on Oil Pollution chaired by an official of the Department of the Environment.

Maritime local and harbour authorities have made contingency plans for the clearance of oil pollution from beaches and immediately off-shore. Some authorities have coordinated their plans or made joint plans; for example, authorities in the south west of Ireland where almost all major pollution incidents occur have made a joint contingency plan for the south west region. Oil pollution officers have been appointed by all maritime local and harbour authorities. The oil pollution officers in local authorities are usually civil engineers who have many other responsibilities. The circular on 'Arrangements for Clearance of Oil Pollution from Coastal Areas' produced by the Department of the Environment lists the names and methods of communicating with these officers. Regional coordinating oil pollution officers have been appointed in most areas.

A Sea Unit Committee, on which various authorities concerned with oil pollution are represented, has been set up to keep arrangements for the clearance of oil at sea under review. Responsibility for sea operations rests with naval officers of the Department of Defence. Authorities who have made joint plans have agreed that there be mutual cooperation in the event of an oil spillage which would over-tax the resources of any one authority. They have also agreed arrangements for reimbursing each other.

6.11.2 Methods of dealing with oil pollution

The Department of the Environment issued circular letters providing general advice on methods of dealing with oil slicks at sea and on the removal of oil from beaches together with information on equipment, on the amount and type of dispersants required, and on the correct usage of dispersants.[62] A number of conferences have been held to bring oil pollution officers up to date on recent developments in methods for dealing with oil pollution. Copies of relevant publications including

Warren Spring Laboratory Oil Pollution Newsletters 3 and 4 and the IMCO *Manual on Oil Pollution—Practical Information on Means of Dealing with Oil Spillages* have been circulated to concerned authorities.

6.11.3 Materials and equipment

Guidelines on suitable materials and equipment have been issued by the Department of the Environment.[63] The guidelines were prepared on the basis that the materials and equipment in any area would be made available if necessary to another area so as to avoid duplication. Appendix II to the guidelines contained information on dispersants and advice on their use. Lists of approved dispersants were given together with the names and addresses of suppliers. Appendix III provided technical information on booms and devices for the mechanical removal of oil from water. The basic requirements of craft suitable for spraying dispersant with inshore Warren Spring Laboratory spraying equipment were given in Appendix IV. The more modern document *Arrangements for Clearance of Oil Pollution from Coastal Areas* (1979) contains information on the nature and location of materials and equipment for dealing with oil pollution.

6.11.4 Communications

The Marine Rescue Coordination Centre (MRCC) at Shannon Airport acts as the communications centre for the receipt of information on oil slicks. This information can be communicated by naval officers or by others, e.g. airplane pilots, who have spotted oil pollution. On receipt of information from air and surface craft, MRCC assesses the likely rate and direction of the oil slick and transmits all relevant information to regional coordinating and/or local oil pollution officers in areas likely to be affected as well as to the responsible officials in the Department of the Environment and the Department of Defence. MRCC keeps notified officers up to date with the situation as it develops.

Notes

1. As amended by OPSA 1977, s. 9.
2. *Federal Steam Navigation Co. Ltd.* v. *Department of Trade and Industry* [1974] 2 A11 E.R. 97.
3. OPSA 1956, s. 3.

4. S.I. No. 353 of 1980.
5. S.I. No. 354 of 1980.
6. OPSA 1977, s. 9.
7. Oil Pollution of the Sea Act 1956 (Exceptions) Regulations 1980, arts 3, 4.
8. OPSA 1977, s. 10(1).
9. *Cosh* v. *Larsen* [1971] 1 Lloyd's Rep. 577.
10. *Ranklin* v. *Da Costa* [1975] 2 All E.R. 303.
11. As amended by OPSA 1977, s. 16.
12. See 4.5 and Local Government (Water Pollution) Act 1977, ss. 3(5) and 4(2).
13. *Arrangements for Clearance of Oil Pollution from Coastal Areas* [1979], Department of the Environment, p. 1.
14. See 6.2.2.
15. See 6.8.
16. *See Liability and Compensation for Oil Pollution Damage* [1979], Department of Trade, London.
17. *Esso Petroleum Co. Ltd* v. *Southport Corporation* [1956] A.C. 218.
18. See Circular ENV. 16/75 of 15 May 1975 and Abecassis: *Oil Pollution from Ships* (1978), London, pp. 235–243.
19. New TOVALOP, clause 1(i).
20. *Ibid.*, clause 1(h).
21. See *Christos Bitas—The Fight Against Pollution* [1978], Department of Trade, London.
22. See Circular ENV. 16/75 of 15 May 1975.
23. See Circular ENV. 19/73 of 25 June 1973.
24. S.I. No. 229 of 1977.
25. Merchant Shipping Act 1894, s. 419(2).
26. Thirty-Eighth Report of the Joint Committee on Secondary Legislation (1977), para. 21.
27. Merchant Shipping (Cargo Ship Construction and Survey) Rules 1967, S.I. No. 99 of 1967.
28. European Communities (Entry Requirements for Tankers) Regulations 1961, S.I. No. 301 of 1981.
29. As amended by OPSA 1977, s. 12.
30. S.I. No. 352 of 1980.
31. As amended by OPSA 1977, s. 13.
32. S.I. No. 119 of 1980.
33. As amended by OPSA 1977, s. 14.
34. See 5.2.
35. Oil Pollution of the Sea Act 1956, s. 24.
36. *Ibid.*, s. 11(2) Comp. Harbours Act 1946, s. 88.
37. *Ibid.*, s. 14.
38. *Ibid.*, s. 15.
39. Dangerous Substances Act 1972, s. 20.
40. *Ibid.*, s. 21(1).
41. *Ibid.*
42. *Ibid.*, s. 32(1) and (3).
43. *Ibid.*, s. 32(2).
44. *Ibid.*, s. 31(1) and (2).
45. *Ibid.*, ss. 32(4), 31(3).
46. *Ibid.*, s. 33(1).
47. *Ibid.*, s. 33(2).
48. *Ibid.*, s. 33(4).
49. *Ibid.*, s. 34(1), 34(2).
50. *Ibid.*, s. 34(3).
51. *Ibid.*, s. 34(4).

52. S.I. No. 311 of 1979.
53. S.I. No. 313 of 1979.
54. S.I. No. 312 of 1979.
55. S.I. No. 314 of 1979.
56. Dangerous Substances Act 1972, s. 63.
57. *Ibid.*, s. 52.
58. *Ibid.*, s. 55.
59. *Ibid.*, s. 36(2).
60. S.I. No. 318 of 1979.
61. See Directory of State Services for 1977, 1978 and 1979.
62. See Circulars ENV. 251/1 of 12 December 1972 and ENV. 16/75 of 15 May 1975.
63. See Circular ENV. 16/75 of 15 May 1975.

7
Discharges to Sewers

Untreated sewage is the second most important cause of water pollution in Ireland,[1] despite the fact that all Irish major cities and urbanised areas are situated on the coast so that most sewage is discharged into coastal waters. The National Survey of Air and Water Pollution published by the IIRS in 1975 listed 32 rivers where serious degradation had occurred and attributed the cause partly or wholly to town sewage in 17 cases. The survey also named nine lakes in a poor or unsatisfactory condition and in three cases attributed the cause partly to town sewage. By 1977 the *Report on Pollution Control* produced by the Inter-departmental Environment Committee stated that remedial works were being planned or in various stages of progress in the twenty towns discharging un-treated or unsatisfactorily treated sewage.[2] But the Committee indicated that future progress must depend upon 'the maintenance of a high level of capital allocation for the sanitary services programme over the next few years'. Priority in capital allocations (which have always been in-adequate) has been given to providing the sanitary infrastructure in new and expanding urban and industrial areas, and many sanitary authorities continue to discharge untreated sewage to waters and to operate sewage works which do not produce effluent to an acceptable standard because they are old, in bad repair or simply overloaded.

7.1 RESPONSIBILITY FOR THE DISPOSAL OF SEWAGE

Control over pollution caused by sewage and discharges to sewers is generally the responsibility of sanitary authorities. The principal Acts governing the powers, duties and responsibilities of sanitary authorities in this area are the Public Health (Ireland) Act 1878, the Public Health Acts Amendment Act 1890, and the Local Government (Water Pollution) Act 1977.

Under the Public Health (Ireland) Act 1878, sewers within their districts, together with the structures ancillary thereto, were with minor exceptions vested in and placed under the control of sanitary authorities.[3] Sanitary authorities were also obliged to 'cause to be made such sewers as may be necessary for effectually draining their district for the purposes of this Act'[4] and were endowed with wide powers to carry out their functions including powers to purchase and make sewers,[5] to alter and discontinue sewers,[6] to dispose of sewage,[7] to agree to the connection of sewers under their control with those of adjoining districts,[8] to deal with land appropriated to sewage purposes,[9] to enter into contracts to carry out their duties[10] and to enforce the provisions of the Public Health Acts 1878 and 1890, relating to nuisances.[11] Authorities were also empowered to make by-laws with respect to, *inter alia*, drainage of buildings, water closets, earth closets, privies, ashpits and cesspools connected with buildings and the keeping of water closets supplied with sufficient water for flushing.[12]

There appears to be some doubt as to whether the duties of sanitary authorities extend to the mandatory reception of trade and industrial effluents. Section 23 of the Public Health (Ireland) Act 1878 provides that once a sewer exists, any individual has the right to empty his drains into it by giving notice to the sanitary authority and on complying with the regulations of the authority as to the manner in which the communication is to be effected. In *Wallace* v. *McCartan*[13] it was held that this right could not be exercised in an unreasonable or negligent manner, or without regard to the rights of third parties. Subject to these common law limitations, there can be no doubt but that the right still exists. The Local Government (Water Pollution) Act 1977, however, ignores its existence and provides in section 16 that a person shall not, after such date as may be fixed for the purpose of that section by the Minister (i.e. 1 January 1980),

> discharge or cause or permit the discharge of any trade effluent or other matter (other than domestic sewage or storm water) to a sewer, except under and in accordance with a licence granted by

the sanitary authority in which the sewer is vested or by which it is controlled.

7.2 DISCHARGES TO SEWERS

7.2.1 Prohibited discharges

As a general rule, Irish legislation does not prohibit the discharge of any particular substance or matter to sewers. However, sections 16 and 17 of the Public Health Amendment Act 1890 are exceptional in that they contain absolute prohibitions on specific discharges. Section 16 provides that it shall be unlawful for any person

> to throw or to suffer to be thrown, or to pass into any sewer of a local authority or any drain communicating therewith, any matter or substance by which the free flow of sewage or surface or storm water may be interfered with, or by which any such sewer or drain may be injured.

The maximum penalty on conviction for an offence under section 16 is £10 plus £1 for each day the offence continues. Section 17 provides that

> any person who causes or permits to enter into any local authority sewer or any drain communicating therewith: (a) any chemical refuse, or (b) any waste steam, condensing water, or other liquid (such liquid being of a higher temperature than 110° Fahrenheit) either alone or in combination with the sewage, causes a nuisance or is dangerous or injurious to health, shall be liable to a penalty not exceeding £10 and to a daily penalty not exceeding £5.

Section 17(3), however, provides that a person shall not be liable to a penalty for an offence under the section until the local authority notifies him of the provisions thereof. Neither is a person liable for an offence committed under section 17(1) before the expiration of 7 days from the service of such notice, it being provided that the local authority shall not be required to give the same person notice more than once.

Section 16(11) of the Local Government (Water Pollution) Act 1977 provides that it is a defence to prosecution under any enactment other than that Act (e.g. sections 16 and 17 above) to prove that the act constituting the offence is authorised under section 16 of that Act. Under section 16(7) of the Water Pollution Act, it is an offence for any person to permit or cause the entry of any polluting matter, including sewage, to any drain or sewer provided solely for the reception or disposal of storm water.

7.2.2 The obligation to obtain a licence for other discharges to sewers

The normal formula for controlling discharges is that employed in section 16(1) of the Local Government (Water Pollution) Act 1977. This section came into force on 1 January 1980. Section 16(1) provides that it shall be unlawful to discharge or cause or permit the discharge of any trade effluent or other matter (other than domestic sewage or storm water) to a sewer except under and in accordance with a licence granted by the appropriate sanitary authority. A 'trade effluent' is defined as

> effluent from any works, apparatus, plant or drainage pipe used for the disposal to waters or to a sewer of any liquid (whether treated or untreated), either with or without particles of matter in suspension therein, which is discharged from premises used for carrying on any trade or industry (including mining) but does not include domestic sewage or storm water.

'Trade' includes agriculture, aquaculture, horticulture and any scientific research or experiment.[14]

Sanitary authorities have been advised by the Department of the Environment that liquid waste arising from agricultural activities does not come within the definition of 'trade' effluent unless it is discharged from a works, apparatus, plant or drainage pipe.[15] Recent guidelines circulated have also suggested that discharges from shops, licensed premises, hotels and other service industry premises when relatively small in volume 'might be regarded as consisting of domestic sewage and consequently not licensable'.[16] A similar suggestion was made in the guidelines with respect to discharges from 'small slaughterhouses operated by individual retailers and provided with adequate grease traps'. In determining whether discharges from so-called 'small trading establishments' ought to be licensed or not, the 'magnitude of the discharge' was to be considered as a deciding factor. It is respectfully submitted that the magnitude of a discharge has little to do with the question of whether an effluent is or is not a 'trade effluent'. The expression is expressly defined in section 1 of the Act as 'effluent . . . discharged from premises used for carrying on any trade or industry (including mining) . . .'. The only effluents from trade premises discharged from any works, apparatus etc. which are not statutorily licensable are domestic sewage or storm water and those which are exempted by the Minister. Domestic effluents and storm water are not statutorily defined but the judicial definition of domestic sewage in the recent English decision *Thames Water* v. *Laundrettes Ltd*[17] as 'liquid from water closets and baths, lavatories and sanitary conveniences' would appear to be acceptable. The correct procedure for ensuring that effluents from small trading establishments and service industry premises are exempt from the provisions of Section 16(1) is to exempt them by regulation under Section 16(10)—not by Departmental circular.

7.2.2.1 EXISTING DISCHARGES

Section 18 of the Local Government (Water Pollution) Act 1977 provides that a person shall not be deemed to have contravened section 16(1) in relation to an existing discharge of trade effluent or other matter made before he is granted or refused a licence under section 16 if, before the relevant date (i.e. 1 January 1979), he applies for such a licence and complies with any regulations under section 19 regarding the furnishing of information to a sanitary authority.

7.2.3 Licence application procedure

The procedures regulating the submission of applications for licences under section 16(1) are contained in Part III of the Local Government (Water Pollution) Regulations 1977.[18] A licence application must be accompanied by:

(i)   such plans, in duplicate, and such other particulars as are necessary to describe the premises, drainage system and any works, apparatus or plant from which effluent or other matter is to be discharged and to identify the point of discharge to the sewer;

(ii)   particulars of the nature, composition, anticipated temperature, volume and rate of discharge of, and the proposed method of any treatment of, the effluent or other matter and the period or periods during which the effluent or other matter is to be discharged;

(iii) in the case of a trade effluent, a general description of the process or activity giving rise to the discharge.[19]

A licence application for an existing discharge must state the above information.[20] An applicant for a licence (other than a licence for an existing discharge) must also furnish such other particulars, including the results of such investigations, as the sanitary authority may reasonably require for consideration of the application.[21] Where an applicant fails or refuses to comply with a requirement of a sanitary authority for other particulars, or results of an investigation within 3 months of being requested to do so, the sanitary authority may carry out the investigation or arrange to have it carried out at the total or partial expense of the

161

applicant.[22] There are no provisions requiring the applicant to publicise his application or requiring a sanitary authority to give public notice of its intention to grant or refuse a licence.

7.2.4 The decision on the application

Section 16 of the Water Pollution Act provides that a sanitary authority may at its discretion refuse to grant a licence, or grant a licence subject to such conditions as it thinks appropriate and specifies in the licence. Conditions attached to a licence may

(i) relate to
  (a) the nature, composition, temperature, volume, rate, method of treatment and location of a discharge and the periods during which a discharge may be made or may not be made;
  (b) the provision and maintenance of meters, gauges, other apparatus, manholes and inspection chambers;
  (c) the taking and analysis of samples, the keeping of records, and furnishing of information to the sanitary authority;

(ii) require (by the payment of a capital sum or an annual charge or both) the defrayment of or contribution towards the cost incurred by the sanitary authority in monitoring, treating or disposing of a discharge;

(iii) specify a date not later than which any condition shall be complied with.[23]

In considering whether to grant a licence, a sanitary authority must have regard to any objectives contained in any relevant water quality management plan.[24] It may not grant a licence in respect of a discharge of a trade effluent which would not comply with any relevant standard prescribed by the Minister for the Environment under section 26 of the Act.[25]

The ordinary individual has no right to inspect documentation lodged with licence applications, or to participate in decision-making on licences. There is no time limit within which a sanitary authority must decide on a licence but 'as soon as may be' after coming to a decision, it must either transmit the licence or give notice of refusal to the applicant and also inform him that section 20 of the Act provides a right of appeal to the Planning Appeals Board where a licence has been refused or granted subject to conditions.[26]

An interesting feature in the Act is the express provision in section

16(4)(*b*) which enables sanitary authorities to require defrayment of or contribution towards the costs of monitoring, treating or disposing of a discharge. Such a practice is relatively rare in Ireland but may be expected to become a common feature of licensing trade and industrial discharges in future. The Minister for the Environment has suggested that, where water pollution is due to a combination of domestic sewage and industrial wastes, a solution to the financial aspects of the problem may lie in the provision of a joint treatment plant (i.e. on the basis of shared costs between local authorities and industry).[27] At present only a few towns operate joint treatment works. A few authorities which impose operating charges calculate them in the proportion which the BOD load and volume discharged bears to overall treatment costs; but a general policy on charges for discharges to sewers has yet to be formulated.

A register must be kept of all licences granted under section 16.[28] Conditions attached to a licence are binding on any person discharging, or causing or permitting the discharge of trade effluent or other matter to which the licence relates.[29] The licence lapses if no discharge of the type authorised by the licence is made within 3 years.[30]

7.2.5 Review of licences

Section 17 of the Water Pollution Act provides that sanitary authorities may review a licence under section 16 at intervals of not less than 3 years from the date of the licence or of the last review of the licence. A licence may be reviewed at any time in five specified circumstances.[31] A sanitary authority must notify a discharger of intention to review a licence.[32] The notice must, *inter alia*, state that written representations relating to the review may be made to the sanitary authority within 1 month.[33] The discharger may be required to submit such plans or other particulars as the sanitary authority considers necessary for the purpose of the review within 3 months.[34] If the required information is not supplied within 3 months, the sanitary authority may complete the review without it.[35] Neither the Act nor the regulations specify a time limit for completing the review. 'As soon as may be' after that, a sanitary authority may amend or delete any condition attached to the licence or attach new conditions or grant a revised licence in substitution for the licence reviewed.[36] Compensation is not payable if the review necessitates expenditure, but the notice of the decision after the review must include a statement that an appeal lies to the Planning Appeals Board under section 20 of the Act.[37]

7.2.6 Appeals to the Planning Appeals Board

Section 20 of the Water Pollution Act provides that

> a person to whom a licence under section 16 has been refused, or granted subject to conditions, may appeal to the Minister in relation to the refusal to grant such a licence, the conditions attached to such a licence or amendment or deletion of conditions attached to or attachment of new conditions to such a licence on review.

The Minister's appellate functions have been transferred to the Planning Appeals Board,[38] which is empowered to deal with appeals in a manner substantially similar to appeals under the Planning Acts.[39] It should be noted that *locus standi* to appeal decisions on licences under sections 16 and 17 is confined to the applicant and the sanitary authority concerned. No individual citizen nor—perhaps more importantly—any other licensee or prospective licensee has the right to appeal a decision. Appeals relating to the grant or refusal of a licence must be brought within 1 month beginning on the date of the grant or refusal; appeals relating to decisions on review must be brought within 1 month of the decision.[40] The Board, after consideration of an appeal, must either 'refuse the appeal or give appropriate directions to the sanitary authority concerned relating to the granting of a licence or the attachment, amendment or deletion of conditions'.[41] Where such directions are given, the authority must, as soon as may be after receiving them, comply with them and notify the holder of the licence of such compliance.[42]

7.2.7 Enforcement

The provisions of the Local Government (Water Pollution) Act 1977, which relate to discharges to sewers, are enforceable by sanitary authorities alone. A person who contravenes section 16(1) or (7) of the Act is liable on summary conviction to a fine not exceeding £250 plus £100 for every day the contravention continues and/or 6 months' imprisonment. The corresponding figures on conviction on indictment are £5000, £500 and 2 years' imprisonment respectively.[43] It is a good defence to a prosecution under any enactment other than the Water Pollution Act that the act constituting the offence is authorised under section 16.[44] As yet, there have been no prosecutions under section 16.

In addition, or as an alternative to their powers to prosecute for breach of subsections 16(1) and 16(7), sanitary authorities may serve a notice on any person contravening these subsections requiring that the contravention be stopped within a specified period and requiring mitigation or remedying of any effects of the contravention within such period and in such manner as may be specified.[45] The sanitary authority may take any steps it considers necessary to prevent the discharge or entry or to mitigate or remedy any effects of the contravention if the person served with the notice does not comply with it within the period specified.[46] Costs are recoverable from the defaulter.[47]

Section 22 of the Act requires sanitary authorities to carry out or arrange for such monitoring and analysis of discharges and receiving waters and empowers them to collect such information as may be necessary for the performance of their function or as may be directed by the Minister for the Environment. Section 23 empowers them to require the submission of specified particulars of discharges to sewers within a specified period. Failure to supply the requested particulars within the time allowed, or the supply of false or misleading particulars, is a offence punishable by a fine not exceeding £100. Section 28 of the Act contains the necessary powers of entry and inspection for the purpose of performing functions under the Act and also provides for the making of regulations governing the taking of samples, carrying out of tests etc.

7.3 COLLECTED SEWAGE

Once in a public sewer, sewage becomes the property of the sanitary authority which is then responsible for treating and disposing of it. Apart from obligations already referred to with respect to providing sewage systems and controlling discharges to sewers, sanitary authorities themselves are empowered under the Public Health (Ireland) Act 1878 to 'receive, store, disinfect, distribute or otherwise dispose of sewage' but the Act,[48] and indeed the common law itself, prohibits sanitary authorities from exercising these powers in such a way as to create a nuisance.

The Public Health (Ireland) Act 1878 obliges sanitary authorities:

(i)   to keep all sewers in repair and to cause to be made such sewers as may be necessary for effectually draining their districts;[49]

(ii)   to purify sewage before discharging it;[50]

(iii)   to cause sewers to be so constructed, covered, ventilated and kept so as not to be a nuisance or injurious to health, and to be properly cleansed and emptied.[51]

The remedy open to a person aggrieved by the failure of a sanitary authority to fulfil the above duties is by way of complaint to the Minister for the Environment under section 15 of the 1878 Act or under section 15 of the Public Health (Ireland) Act 1896, which give the individual statutory rights to complain to the Local Government Board (now the Department of the Environment) of the failure of sanitary authorities to carry out their duties. The 1896 Act is particularly useful in that it provides that the Department, (a) on receipt of a complaint that a sanitary authority has defaulted in providing its district with sufficient sewers, or in the maintenance of existing sewers, or that a local authority has made default in enforcing the provisions in the Public Health (Ireland) Acts 1878 to 1890, which it is its duty to enforce, and (b) on being satisfied that the authority has been guilty of the alleged default, is obliged to make an order limiting a time for the performance of the duty in the matter complained of. If the duty is not performed in the time specified, the order may be enforced by mandamus or the Department itself may act in default. An individual who has suffered injury to his private rights may also sue the sanitary authority for breach of statutory duty.[52]

7.4 DISCHARGES FROM SEWERS OR SEWAGE TREATMENT PLANTS

The licensing requirements of section 4 of the Local Government (Water Pollution) Act 1977 do not apply to discharges 'from a sewer'.[53] Instead, section 26(1) of that Act provides that the Minister for the Environment may, after consulting the Ministers for Fisheries, Industry and Commerce and any other Minister who appears to him to be interested, and the Water Pollution Council, prescribe for the purposes of the Act quality standards for, *inter alia*, waters and sewage effluents and standards in relation to the methods of treatment of such effluents. Section 26(3) specifically states that where regulations under section 26 relate to sewage effluent from a sewer or to waters to which sewage effluent from a sewer discharges, it shall be the duty of the sanitary authority in which the sewer is vested or by which it is controlled to take steps as soon as practicable to ensure that the sewage effluent complies with, or does not result in the waters to which the effluent is discharged not complying with, any relevant standard prescribed under that section.

To date, the Minister has not prescribed any standards under section 26. Indeed, it would be politically unrealistic for him to prescribe mandatory standards unless he also provided the financial resources which would enable sanitary authorities to comply with them. In practice sanitary authorities seek to achieve the standards recommended by the

UK Royal Commission on Sewage Disposal for sewage effluents discharged to inland waters. The Water Quality Guidelines circulated to sanitary authorities in 1979 recommend standards for water quality and sewage effluents but these are not mandatory.[54]

7.5 ENFORCEMENT

There is very little information available as to the extent to which provisions in the Acts described in this section are enforced. Matters relating to sewage belong almost exclusively to the public domain. Sanitary authorities are prosecuted for causing water pollution from time to time,[55] but the solution to their problems and that of pollution caused by their failure to deal with sewage properly is a financial rather than a legal one.

Notes

1. *Report on Pollution Control,* p. 6.
2. *Ibid.,* p. 14.
3. Public Health (Ireland) Act 1878, s. 15.
4. *Ibid.,* s. 17.
5. *Ibid.,* ss. 16, 17.
6. *Ibid.,* s. 20.
7. *Ibid.,* s. 30.
8. *Ibid.,* s. 31.
9. *Ibid.,* s. 32.
10. *Ibid.,* s. 200.
11. *Ibid.,* ss. 107–127.
12. *Ibid.,* s. 41 as extended by s. 23 of the Public Health Acts Amendment Act 1890.
13. *Wallace* v. *McCartan* [1917] 1 I.R. 377. See also *Dublin County Council* v. *Short* H.C. 21.12.1981 (unreported) on the right to discharge to sewers.
14. Local Government (Water Pollution) Act 1977, s. 1.
15. Circular ENV. 18/77 of 19 September 1977.
16. Circular ENV. 2/80 of 24 April 1980.
17. *Thames Water* v. *Laundrettes Ltd* [1980] 1 W.L.R. 700, p. 706.
18. Local Government (Water Pollution) Regulations 1977 (S.I. No. 108 of 1978).
19. *Ibid.,* art. 17(1).
20. *Ibid.,* art. 17(2).
21. *Ibid.,* art. 17(3).
22. *Ibid.,* art. 18.
23. Local Government (Water Pollution) Act 1977, s. 16(4).
24. *Ibid.,* s. 16(2). See also 4.4.5.
25. *Ibid.,* s. 16(3). The Minister has not yet prescribed any standards under section 26.
26. Local Government (Water Pollution) Regulations 1977, art. 20.
27. Circular ENV. 2/80 of 24 April 1980.
28. Local Government (Water Pollution) Act 1977, s. 9.
29. *Ibid.,* s. 16(5).

30. *Ibid.*, s. 16(6).
31. *Ibid.*, s. 17(1) and 17(3).
32. Local Government (Water Pollution) Regulations 1977, art. 22(1).
33. *Ibid.*, art. 22(2).
34. *Ibid.*, art. 23(1).
35. *Ibid.*, art. 23(2).
36. Local Government (Water Pollution) Act 1977, s. 17(2).
37. Local Government (Water Pollution) Regulations 1977, art. 24.
38. Local Government (Water Pollution) Act 1977 (Transfer of Appeals) Order 1978 (S.I. No. 96 of 1978).
39. *Ibid.* See also 2.4.5.
40. Local Government (Water Pollution) Regulations 1977, art. 26.
41. Local Government (Water Pollution) Act 1977, s. 20.
42. Local Government (Water Pollution) Regulations 1977, art. 35.
43. Local Government (Water Pollution) Act 1977, s. 16(8).
44. *Ibid.*, s. 16(11).
45. *Ibid.*, s. 16(13).
46. *Ibid.*, s. 16(14).
47. *Ibid.*
48. Public Health (Ireland) Act 1878, s. 30.
49. Public Health (Ireland) Act 1878, s. 17.
50. *Ibid.*, s. 19.
51. *Ibid.*, s. 21.
52. See *Buckley* v. *Kerry County Council and Kerry Board of Health* (1928) 62 I.L.T.R. 127.
53. Local Government (Water Pollution) Act 1977, s. 4(2)(*b*), but the definition of sewer in this Act is a limited one.
54. *Memorandum No. 1 on Water Quality Guidelines* (1979), pp. 10, 35.
55. See 4.2.

8
Waste on Land

An examination of waste disposal laws and practices in Ireland results in an alarming picture of the inadequacy of existing legislation and of arrangements made for waste disposal on land.

Until 1979 the main sources of law on waste disposal were the Public Health (Ireland) Act 1878 and the Public Health Acts Amendment Act 1907. Provisions in other Acts, notably the Petroleum and Other Minerals Development Acts 1940 and 1960, the Local Government (Planning and Development) Acts 1963 and 1976, and the Local Government (Water Pollution) Act 1977, also operate to control improper waste disposal. In 1979 the European Communities (Waste) Regulations 1979, the European Communities (Waste) (No. 2) Regulations 1979 and the European Communities (Toxic and Dangerous Wastes) Regulations 1982 were enacted to ensure compliance with EEC obligations but these can in no way be described as providing a satisfactory system for the regulation of waste disposal in Ireland.

Local authorities are now responsible for the planning, authorisation and supervision of waste operations in their areas. But their obligations with respect to waste collection are limited to the collection of household and trade wastes only, and they are not specifically obliged to provide for the disposal of any wastes other than household wastes. In practice many local authorities dispose of trade waste or allow it to be disposed of on their dumps. The collection and disposal of industrial wastes is generally left to private concerns which were not regulated by law until 1979. Many of these concerns operate in an irregular manner. There is no official toxic waste dump in the country. Farm wastes are generally disposed of according to good agricultural practices and the Department of Agriculture actively promotes these practices through the Agricultural Advisory Services.[1]

There are, as yet, no legislative provisions specifically dealing with such matters as the operation of waste disposal sites, and the manner in which wastes are to be treated prior to their disposal on land. Neither are there any legislative incentives aimed at encouraging citizen participation in decision-making on waste disposal matters, at reducing at source the quantities of wastes generated, or at encouraging the beneficial use of waste products. The Department of the Environment has been preparing comprehensive legislation on waste disposal for several years.

8.1 WASTE DISPOSAL SITES

8.1.1 Public waste disposal sites

Waste disposal is carried out by local authorities or private concerns. Statutory responsibility for waste disposal rests with sanitary authorities under the Local Government (Sanitary Services) Acts 1878–1964, and with local authorities under the European Communities (Waste) Regulations 1979[2] and the European Communities (Waste) (No. 2) Regulations 1979.[3] Section 55 of the Public Health (Ireland) Act 1878 obliges sanitary authorities to provide 'fit buildings and places' for the deposit of any matters collected by them in pursuance of Part II of that Act. This obligation did not extend to the provision of places for the deposit of trade or industrial waste as these were not dealt with in Part II of the Act, but section 48 of the Public Health Acts Amendment Acts 1907 provides that local authorities must *remove* trade refuse (other than sludge) if required to do so by the owner or occupier of any premises. The Act does not, however, deal with the *disposal* of trade wastes. Sanitary authorities, as local authorities, have rather inadequate powers—but not duties—to provide places for the disposal of waste under the Local Government (Planning and Development) Acts 1963 and 1976. Under the 1963 Act, planning (i.e. local) authorities are obliged to make development plans for their areas. Among the objectives which may be indicated in development plans are provisions for 'prohibiting, regulating and controlling the deposit and disposal of waste materials and refuse'.[4] Planning authorities are obliged to take such steps as are necessary for securing the objectives which may be contained in the provisions of their development plans[5] and may therefore provide sites for waste disposal. The selection and designation of a waste disposal site, if contained in a draft development plan or variation or amendment thereof, is subject to the controls (which require that the public be given an opportunity to participate in the decision-making process) in Part III of the 1963 Act.[6] But the development and operation of such a site, if carried out by a local authority in its own area, constitutes 'exempted' development and need not be authorised under Part IV of the 1963 Act.[7] It is, however, a general principle of the common law and sometimes also an express statutory requirement that public authorities may not carry out their duties in such a way as to cause a nuisance.[8] Thus a public dump may not be operated so as to occasion a nuisance. The Minister for the Environment has advised local authorities in circular letters on site selection and on the operation of their waste disposal sites.[9]

Under article 4(2) of the European Communities (Waste) Regulations 1979, local authorities are obliged to prepare a plan indicating, *inter alia*, suitable waste disposal sites. The Minister for the Environment has required the submission of draft waste disposal plans to his Department by 30 June 1980 and has provided local authorities with an outline plan to promote standardisation in plan content.[10] Plans are required to contain a map indicating local authority sites for (a) household waste only, (b) general waste (excluding hazardous waste), (c) disposal facilities for hazardous wastes, and (d) local authority waste oil depots. Under article 3 of the European Communities (Toxic and Dangerous Waste) Regulations 1982, local authorities are required to prepare toxic and dangerous waste plans indicating, *inter alia*, suitable disposal sites for wastes regulated under EEC Directive 78/319/EEC on Toxic and Dangerous Waste but local authorities themselves are not expressly required to provide such sites.

There is no express legal obligation on local or sanitary authorities to provide places for the disposal of any wastes other than what might loosely be termed domestic wastes. There are, as has been illustrated, implied powers for local authorities to provide such places. In view of the difficulties—legal and political—experienced by Cork County Council in trying to select and provide a site for a toxic waste dump, specific enabling legislation on these matters would appear to be necessary.[11]

In June 1981 the Minister for the Environment notified local authorities of a new waste disposal strategy which he proposed adopting in order to ensure that adequate provision is made for the proper disposal of wastes of all kinds. The main proposals are:

(i)   the establishment of a central depot and possible treatment facilities for industrial and other wastes for which acceptable alternative means of disposal are not available;

(ii)   development of a network of disposal tips around the country;

(iii)   financial support to local authorities for upgrading existing tips and acquiring and developing new tips for co-disposal use;

(iv)   continued consideration of appropriate sea disposal services, and use, where appropriate, of local authority sewage treatment plants for liquid waste suitable for disposal by this means;

(v) general extension and improvement of tips to meet local authority needs for the disposal of trade and domestic wastes to acceptable standards;

(vi) strengthened powers to local authorities to control waste operations and prosecute offenders;

(vii) the undertaking of a new planning activity by local authorities to promote proper waste disposal and the implementation of the strategy as it affects their areas.[11a]

8.1.2 Private waste disposal sites

The development and operation of a waste disposal site by any person other than a local authority is subject to control under section 3 of the Local Government (Planning and Development) Act 1963, which expressly provides that where land is used for 'the deposit of bodies or other parts of vehicles, old metal, mining or industrial waste, builder's waste, rubble or debris', the use of the land shall be taken as having materially changed. Planning permission under Part IV of the Act is therefore necessary, and conditions regulating the development and operation of the site may be attached to any planning permission granted.[12] Unfortunately, the vast majority of private waste disposal sites are not controlled under the Planning Acts either because they were in existence before October 1964 or because local authorities have chosen to ignore their existence.

If land is being used for waste disposal purposes without the required planning permission or in breach of a condition attached to a permission, a local authority or a private individual may have recourse to enforcement powers under the Planning Acts 1963 and 1976.[13] In addition, section 107 of the Public Health (Ireland) Act 1878 states that 'any accumulation or deposit which is a nuisance or injurious to health' shall be deemed to be a statutory nuisance. Local authorities are obliged under sections 110 and 111 of the Act to abate such nuisances on receipt of complaints about them from specified persons. A private individual is also empowered by section 121 of the 1878 Act to prosecute in respect of a statutory nuisance.[14]

Under article 5(3) of the European Communities (Waste) Regulations 1979, it is an offence for any person, other than a public waste collector,[15] to carry out the treating, storing or tipping of waste on behalf of another person without a permit from his local authority or in a manner contravening the terms of such a permit. The operation of a private dump on a commerical basis must thus be authorised under the regulations as well as under the Planning Acts. The penalty on summary conviction for operating a private dump without a permit is a fine not exceeding £600 and/or 6 months' imprisonment.[16] Other penalties may be imposed under the Planning Acts. A permit must specify the type and quantity of waste to which it applies and must be conditional on compliance with any general technical requirements and precautions it specifies and on the making available to the local authority of such information as it requests to the origin, destination, treatment, type and quantity of the waste.[17]

Article 5 of the European Communities (Toxic and Dangerous Waste) Regulations 1982, forbids any person other than a local authority to carry out the storage, treatment or deposit of a toxic or dangerous waste other than under and in accordance with (i) the terms of a local authority permit, and (ii) the requirements in articles 6, 7 and 8 of the Regulations. The maximum penalty for breach of any of the regulations is £1000 and/or 6 months' imprisonment.

8.2 WASTES IN GENERAL

The most comprehensive controls over wastes in general, as distinct from specific wastes, are those contained in the European Communities (Waste) Regulations 1979, which were enacted specifically to give effect to EEC Directive 75/442/EEC on waste. The regulations came into operation on 1 April 1980 but have not yet been fully implemented. Local authorities are charged with responsibility for the 'planning, organisation, authorisation and supervision of waste operations in their area',[18] and are required to prepare waste disposal plans. The Minister for the Environment has required the preparation of plans by 30 June 1980 and has requested local authorities to consider the preparation of joint waste disposal plans in areas which are closely interdependent in regard to the movement of waste.[19] A holder of waste must not permit its disposal by any person other than (a) a public waste collector or (b) a permit holder. If he disposes of waste himself he must not do so in such a way as would endanger human health or harm the environment.[20] A person other than a public waste collector must have a local authority permit for creating, storing or tipping waste if doing this on behalf of another person.[21] Permits may be limited to certain wastes or certain quantities of wastes and must be conditional on compliance with any general technical requirements and precautions specified and on making specified information available to the local authority when requested.[22] Registers of waste operations must be kept showing the type and quantities of all wastes handled and their origin, treatment and destination.[23] Authorised persons have been given the necessary powers of entry and inspection.[24] The penalty on summary conviction for operating without a permit or in contravention of the terms therein is a fine not exceeding £600 and/or 6 months' imprisonment.[25] Penalties (£250 on summary conviction) are also provided for failure to keep or produce a proper register of waste operations or to give relevant information to an authorised person and for obstructing or interfering with such a person.[26]

The regulations do not apply to:

(i)   radioactive waste;

(ii)   waste resulting from prospecting, extraction, treatment and storage of mineral resources and the working of quarries;

(iii) animal carcasses and agricultural waste comprising faecal matter or substances used in farming;

(iv)   waste waters (but not including waste in liquid form);

(v)   gaseous effluents emitted into the atmosphere.[27]

8.3 SPECIFIC WASTES

8.3.1 Waste oils

The European Communities (Waste) Regulations 1979 were enacted, *inter alia*, to implement EEC Directive 75/439/EEC on waste oils.[28] The total lubricating oil market in Ireland in 1977 was 8.7 million gallons and the maximum recoverable quantity of waste oils was about 2.7 million gallons. Local authorities have set up a number of waste oil depots in association with disposal dumps but there are still many areas where waste oils are not collected or disposed of in a manner envisaged by the Directive. Some waste oils are collected by private contractors and are recycled and used for horticultural heating purposes.

8.3.2 Polychlorinated biphenyls and polychlorinated terphenyls

EEC Directive 76/403/EEC on the disposal of polychlorinated biphenyls and terphenyls[29] has been implemented by the European Communities (Waste) (No. 2) Regulations 1979, the purpose of which is to ensure the safe disposal of waste PCB, including PCB contained in objects or equipment no longer capable of being used. The Electricity Supply Board (ESB)—a semi-State body—has been designated as the authority authorised to dispose of PCB.[30] Holders of waste PCB must notify the ESB and make it available for disposal by them.[31] The ESB must maintain a register of notifications and of waste held or disposed of and must furnish information required by the Minister for the Environment.[32] It is an offence punishable on summary conviction by a fine not exceeding £600 and/or 6 months' imprisonment for any person to discharge, dump or tip PCB in a way which would endanger human health or harm the environment. The regulations are enforceable by the Minister for the Environment.[33] Further controls over the use of PCBs are contained in the European Communities (Dangerous Substances and Preparations) (Marketing and Use) Regulations 1979,[34] and the European Communities (Dangerous Substances) (Classification, Packaging and Labelling) Regulations 1979.[35]

8.3.3 Waste from the titanium dioxide industry

It has not been found necessary to implement EEC Directive 76/176/EEC on waste from the titanium dioxide industry because this product is not manufactured in Ireland.[36]

8.3.4 Toxic and dangerous wastes[37]

In March 1982 the Minister for the Environment made the European Communities (Toxic and Dangerous Waste) Regulations 1982, to give effect to Directive 78/319/EEC. The regulations will not come into force until 1 January 1983. 'Toxic and dangerous wastes' as defined in the Regulations means any waste containing or contaminated by the substances or materials listed in the Annex to the Directive of such a nature, in such quantities or in such concentrations as to constitute a risk to health or the environment.

There are twenty-seven listed categories of toxic and dangerous wastes. Local authorities in the Counties and County Boroughs and in Dun Laoghaire are charged with responsibility for the planning and control of toxic waste disposal in their areas, for the preparation of disposal plans, and for the authorisation of the storage, treatment and depositing of such wastes. A local authority permit, with or without controlling conditions attached, must be obtained by any person storing, treating or depositing a listed waste. A person producing or holding a listed waste is obliged to have it dealt with by a permit holder. Requirements are prescribed relating to the handling of listed wastes and registers must be maintained by permit holders and by any person producing or holding such wastes. Persons transporting a listed waste must ensure that it is accompanied by a consignment note in a prescribed form containing details of the waste and precautions to be taken. Appropriate inspection and enforcement powers are given to persons authorised by the Minister for the Environment or a local authority. The penalty on summary conviction for operating without a permit or in contravention of the terms therein is a fine not exceeding £1000 and/or six months' imprisonment. Similar penalties are provided for failure to keep or produce a proper register of waste operations, for failure to give relevant information to an authorised person and for obstructing or interfering with such a person.

The Minister for the Environment also announced in March 1982 that a 'national centre and disposal service for certain potentially hazardous wastes' was to be developed at Baldonnell outside Dublin. The centre is to provide a service for producers of hazardous wastes who do not have alternative satisfactory means of disposal, facilities for the reception and storage of the wastes, and export arrangements for their treatment or disposal at authorised facilities abroad. Treatment facilities may be provided at a later stage if necessary.

8.3.5 Trade wastes

Section 128 of the Public Health (Ireland) Act 1878 states that the written consent of the appropriate local authority must be obtained before an 'offensive' trade can be established in an urban area. An offensive trade includes the trade of blood-boiler, bone-boiler, fellmonger, soap-boiler, tallow melter, tripe-boiler, gut manufacturer or any other noxious or offensive trade, business or manufacture. The penalty on conviction for breach of section 128 is a maximum fine of £50 plus £2 each day for a continuing offence. Local authorities are not empowered to attach conditions to consents given but section 129 of the Act empowers them to make by-laws with respect to offensive trades 'in order to prevent or diminish the noxious or injurious effects thereof'. These provisions are of little practical importance nowadays and have to a large extent been superseded by requirements in planning legislation. The Public Health Acts Amendment Act 1907 provides for the removal of trade refuse by local authorities subject to a reasonable payment for the service. Trade wastes are also regulated under the European Communities (Waste) Regulations 1979.

8.3.6 Domestic refuse

The Public Health (Ireland) Act 1878 provides that a sanitary authority may, and if required by the Minister for the Environment must, undertake or contract for the removal of house refuse.[38] Once such service has been undertaken or contracted for, a sanitary authority will be guilty of an offence if, having been required by notice from the occupier of any house to remove such refuse within 7 days, it fails without reasonable excuse to perform or have that service performed.[39] Where a sanitary authority does not contract for or undertake the removal of house refuse, it may make by-laws imposing the duty of such removal on the occupier of premises.[40] Sanitary authorities are also obliged to provide, if necessary, receptacles for the temporary deposit and collection of dust, ashes and rubbish provided that no nuisance is created by the exercise of this power.[41] The deposit of domestic waste may also be controlled under local authority by-laws[42] and under the European Communities (Waste) Regulations 1979.

8.3.7 Litter

Section 52 of the Local Government (Planning and Development) Act 1963 provides that it shall be an offence for any person to throw down, place or leave in or on any public place:

(a) any food remnants, orange peel, banana skin or other organic matter (whether waste or animal), or,

(b) any rubble, old metal, glass, china, earthenware, tin, carton, paper, rags or other rubbish,

so as to create a nuisance.

The penalty is a fine not exceeding £10. This offence may be prosecuted by the planning authority in whose area the offence is committed.

Local authorities are also empowered to make anti-litter by-laws. Several local authorities have done so. Sanitary authorities have powers to provide receptacles for the temporary deposit of rubbish.[43]

In 1979, the Minister for the Environment instructed the Environment Council to report to him on solutions to the litter problem. A report, *Litter and the Environment*, was published in 1980. In March 1981 a Litter Bill was introduced in the Dail but was not passed by the Oireachtas before the dissolution of the Dail in June 1981. This Bill, if enacted in its present form, will considerably improve local authority controls over litter, abandoned vehicles and metal scrap. It will also repeal section 52 of the Planning Act above.

8.3.8 Street cleaning

Section 52 of the Public Health (Ireland) Act 1878 provides that local authorities may, and if required by the Minister for the Environment must, undertake or contract for the proper watering of streets in their district. Where a sanitary authority does not undertake or contract for the cleansing of streets adjoining premises, section 54 of the Act provides that they may make by-laws imposing such a duty on the occupier of the premises. Similar powers are given to local authorities in the Litter Bill 1981. Furthermore, the Bill provides that creating litter in a public place or in a private place visible from a public place will be an offence punishable on summary conviction by a maximum fine of £500.

8.3.9 Abandoned vehicles and scrap metal

The Road Traffic (Removal, Storage and Disposal of Vehicles) Regulation 1971[44] empowers road authorities[45] to remove, store and dispose of a vehicle which has been abandoned on a public road or in a car park provided under section 101 of the Road Traffic Act 1961. This power includes power to make arrangements with any person for the removal and storage of such vehicles. Section 97 of the 1961 Act defines a vehicle as including part of a vehicle or article designed as a vehicle but not at the time capable of functioning as a vehicle.

Local authorities have been requested to provide special dumps in central locations for old vehicles and 'worn out bulky domestic equipment'. They have also been encouraged to cooperate with private companies engaged in the reclamation of scrap metal.[46] Under the terms of the

Litter Bill 1981, local authorities are given wide powers of control over abandoned vehicles and metal scrap including powers to remove or require the removal of abandoned vehicles from any land in their areas and to provide or secure the provision of places for abandoning vehicles and metal scrap.

The use of privately owned land for the dumping of vehicles or vehicle parts must be permitted under Part IV of the Local Government (Planning and Development) Act 1963.[47]

8.3.10 Mining waste

The Minerals Development Act 1940 and the Petroleum and Other Minerals Development Act 1960 contain a number of provisions by which extensive controls over waste from mines may be exercised. These controls include:

(i)   powers to attach conditions to licences, leases, permits and permissions granted under the said Acts;

(ii)   provisions requiring the payment of compensation for damage caused to the surface of land or to mineral deposits or water supplies, or for causing a nuisance, by mining or petroleum operations;

(iii)   provisions requiring that operations be carried on in such a manner as not to interfere with the amenities of the locality near the mines, etc.;

(iv)   provisions empowering the Minister for Industry and Commerce to make regulations in respect of, *inter alia*, the disposal of waste products.

Furthermore, mineral extraction and ancillary operations, such as the erection of buildings and tipping of waste, come under planning control.

Mining waste is exempt from the provisions of the European Communities (Waste) Regulations 1979, and the European Communities (Toxic and Dangerous Waste) Regulations 1982.

8.3.11 Miscellaneous

Provisions in various statutes and other less formal methods of pollution control may operate to prevent the deposit of wastes or specific kinds of wastes in certain places. The deposit of domestic wastes from vehicles or trailers to roads is controlled under the Road Traffic (Construction Equipment and Use of Vehicles) Regulations 1963.[48] The deposit of waste on the foreshore is regulated under the Foreshore Act 1933.[49] The deposit of alkali waste on land is regulated under the Alkali etc. Works Regulation Act 1906.[50] Other controls in water and air pollution legislation may operate indirectly to prevent the deposit of waste on land.

8.4 CONTROLS OVER WASTE TREATMENT BEFORE DISPOSAL AND METHODS OF WASTE DISPOSAL

8.4.1 Disposal by public authorities

It is a general requirement of the common law and sometimes also an express statutory requirement that local and sanitary authorities in carrying out their duties do so in such a way as not to create a nuisance. This requirement may operate as a general control over the method in which waste is treated before disposal and the methods by which waste is disposed of. Otherwise the only legislative provisions on the treatment of waste by public authorities before disposal and on waste disposal methods are contained in the European Communities (Waste) Regulations 1979, which merely provide that a waste disposal plan 'may include measures to encourage rationalisation of the collection, sorting and treatment of waste'.[51] The Minister for the Environment has indicated in circulars that he is less than satisfied with standards observed by many local and sanitary authorities in their waste disposal operations and has issued several circulars recommending treatment methods.[52] The most common recommended method of waste disposal is controlled tipping whereby refuse is dumped in an inoffensive manner and land is reclaimed for amenity or agricultural use.[53] Waste is ideally compacted before being deposited and covered with soil or other suitable inert material.

Local authorities have been advised to cooperate with private concerns involved in recycling scrap metal[54] and to include measures to promote or support recycling of wastes and the recovery of waste materials in their waste disposal plans.[55] There are no other legal or administrative requirements on recycling wastes.

Outlines of a draft aquifer protection policy which, if adopted by the

Department of the Environment, will result in changes in the methods by which waste disposal sites are selected and in existing methods of waste disposal, were prepared by the Geological Survey Office in 1979.[56]

8.4.2 Disposal by others

Various penalties provided in respect of improper methods of waste disposal by persons other than local authorities have been described above. Controls over the methods of treatment of wastes before disposal and of waste disposal methods may be imposed by conditions attached to planning permissions or approvals granted under the Local Government (Planning and Development) Acts 1963 and 1976.[57] Under the European Communities (Waste) Regulations 1979, the permits which must be obtained from local authorities before a person may engage in treating, tipping or storing wastes on behalf of others must be conditional 'on compliance with any general technical requirements and precautions' specified in the permit.[58] These requirements and precautions may require the treatment of wastes before disposal or the adoption of particular waste disposal methods. A holder of waste, if he does not dispose of it himself, must give it to a public waste collector or to a person permitted to dispose of waste.[59] If he does dispose of it himself, he is prohibited from doing so in a manner which would endanger human health or harm the environment.[60] Requirements relating to the preparations (e.g. sorting, packaging) of trade and household wastes for collection by local authorities could be made in by-laws.[61] The European Communities (Toxic and Dangerous Waste) Regulations 1982 prescribe disposal requirements in respect of the toxic and dangerous wastes regulated therein. The IDA may also attach conditions with respect to many aspects of waste disposal to grant-aids.[62]

8.5 ENFORCEMENT

There are no published statistics on the extent to which the laws discussed in this section are enforced. Local authorities do not employ personnel specially to enforce waste control laws. In the last year a few prosecutions for the improper dumping of toxic wastes have been initiated under the Local Government (Water Pollution) Act 1977, but no attempt has yet been made to enforce the European Communities (Waste) Regulations 1979.

181

8.6 INDIVIDUAL RIGHTS

Statutes regulating the activities of public (as distinct from private) bodies rarely contain provisions on individual rights. The only statutory rights the individual enjoys with respect to participation in, and enforcement of, waste laws are those contained in the Public Health (Ireland) Act 1878 (with respect to the enforcement of statutory nuisances),[63] and in Part III of the Local Government (Planning and Development) Act 1963.[64] Under the terms of the Litter Bill 1981 the occupier of land on which a vehicle has been abandoned is empowered to prosecute the person responsible. In addition, any citizen whose rights are infringed by the improper disposal of waste may be able to sue in negligence, nuisance, trespass or under the Rule in *Rylands* v. *Fletcher*.[65]

Notes

1. See 1.4.4. In July 1980 these functions were assumed by An Comhairle Oiliuna Talmhaiochta.
2. S.I. No. 390 of 1979.
3. S.I. No. 388 of 1979.
4. Local Government (Planning and Development) Act 1963, Third Schedule, Clause 11.
5. *Ibid.*, s. 22.
6. See 2.1.4.
7. See 2.4.1.
8. See Public Health (Ireland) Act 1878, s. 55.
9. Circular ENV. 2/74 of 8 January 1974.
10. Circulars ENV. 13/76 of 9 July 1976 and ENV. 10/78 of 19 April 1978.
11. See *Irish Times*, 16 April 1979 and 24 June 1979.
11a. Circular ENV 5/1981 of 4 June 1981.
12. See 2.4.4.
13. See 2.4.8.
14. See 3.1 and 3.11.
15. A public waste collector is the local or sanitary authority.
16. European Communities (Waste) Regulations 1979, art. 5(4).
17. *Ibid.*, art. 5(2).
18. *Ibid.*, art. 4.
19. Circular ENV. 10/78 of 19 April 1978.
20. European Communities (Waste) Regulations 1979, art. 3(1).
21. *Ibid.*, art. 5(3).
22. *Ibid.*, art. 5(2).
23. *Ibid.*, art. 6(1).
24. *Ibid.*, art. 6(2).
25. *Ibid.*, art. 3(2).
26. *Ibid.*, art. 6(3).
27. *Ibid.*, art. 8.
28. OJ L 194/23, 25 July 1975.
29. OJ L 108/41, 26 April 1976.
30. European Communities (Waste) (No. 2) Regulations 1979, art. 1.
31. *Ibid.*, art. 4.

32. *Ibid.*, art. 7.
33. *Ibid.*, art. 5.
34. S.I. No. 382 of 1979.
35. S.I. No. 383 of 1979.
36. OJ L 54, 25 February 1978.
37. See *Today's and Tomorrow's Waste*, Proceedings of Seminar on Waste (1979), National Board of Science and Technology, and EEC Directive 78/319/EEC, O.J. No. L 84/43, 31.3.1978.
38. Public Health (Ireland) Act 1878, s. 52.
39. *Ibid.*, s. 53.
40. *Ibid.*, s. 54.
41. *Ibid.*, s. 55.
42. Municipal Corporations (Ireland) Act 1840, s. 125; Local Government (Ireland) Act 1898, s. 16.
43. Public Health (Ireland) Act 1878, s. 55.
44. S.I. No. 5 of 1971.
45. Road authority means (a) the council of a county, (b) the corporation of a county or other borough, or (c) the council of an urban district (Road Traffic Act 1961, s. 3).
46. Circular ENV. 6/75 of 1 March 1975.
47. Local Government (Planning and Development) Act 1963, s. 3.
48. Road Traffic (Construction Equipment and Use of Vehicles) Regulations 1963, s. 33.
49. Foreshore Act 1933, ss. 13, 14.
50. Alkali etc. Works Regulation Act 1906, s. 4.
51. European Communities (Waste) Regulations 1979, art. 4(3).
52. Circular ENV. 6/75 of 14 March 1975.
53. Circular ENV. 2/74 of 8 January 1974 and Circular ENV. 5/1981 of 4 June 1981.
54. Circular ENV. 6/75 of 14 March 1975.
55. Circular ENV. 10/78 of 19 April 1978.
56. See Wright, G. R. and Daly, D., 'The Hydrogeological Aspects of Tip Site Selection' and 'Groundwater Pollution and Protection', Papers at Conference on Planning for Waste Disposal (1979), An Foras Forbartha.
57. See 2.4.4.
58. European Communities (Waste) Regulations 1979, art. 5(2).
59. *Ibid.*, art. 3(1)(*a*).
60. *Ibid.*, art. 3(1)(*b*).
61. Municipal Corporations (Ireland) Act 1840, s. 125; Local Government (Ireland) Act 1898, s. 16.
62. See 1.5.3.
63. See 3.1 and 3.11.
64. See 2.1.
65. See 1.2.

9
Noise and Vibration

There are very few statutory controls over noise and vibration in Ireland. There are no national noise level standards. Neither are there any national standards for noise emissions from construction plant and equipment, motorcycles, household goods nor many other products which generate excess noise. EEC standards on permissible sound levels from some products have been adopted.[1]

Control over noise, such as it is, is exercised primarily through conditions attached to planning permissions, to authorisations of various kinds for certain activities, and to grants which may be available from the Industrial Development Authority.[2]

Local authorities are primarily responsible for the enforcement of noise controls though this is an area where greater reliance than usual is placed on individual enforcement.

9.1 CONTROLS OVER LAND USE

Controls over the siting and location of developments which might create or be subject to excess noise or vibration may be exercised by land-use zoning in development plans.[3] Section 19 of the Local Government (Planning and Development) Act 1963 specifically declares that all plans in county boroughs, urban districts and scheduled towns shall include objectives restricting particular areas for particular purposes. The Third Schedule of the Act, which specifies objectives which may be indicated in development plans, has a special section on roads and traffic and a section on structures which specifically enables the regulation and control of 'building lines, coverage and the space about dwellings and other structures . . .'.[4] Local authorities in their capacities as planning and road authorities use the UK Department of the Environment publication

Calculation of Road Traffic Noise[5] when predicting noise levels from proposed new roads.[6] This publication permits the calculation of L_{10} noise level for the period 06.00 to 24.00 hours for various road and traffic conditions. UK recommendations on tolerable noise levels are also followed.

Under the Air Navigation and Transport Act 1950, the Minister for Transport is also empowered to make orders imposing restrictions prohibiting building or the erection of structures exceeding a specified height in areas near aerodromes.[7]

Conditions relating to noise and vibration may also be attached to planning permissions or approvals. Section 26(2) of the Local Government (Planning and Development) Act 1963, as amended,[8] expressly provides that conditions attachable to planning permissions may include

conditions for requiring the taking of measures to reduce or prevent:

(i)   the emission of any noise or vibration from any structure comprised in the development authorised by the permission which might give reasonable cause for annoyance either to persons in any premises in the neighbourhood of the development or to persons lawfully using any public place in that neighbourhood, or

(ii)   the intrusion of any noise or vibration which might give reasonable cause for annoyance to any persons lawfully occupying any such structure.

Planning authorities have been advised that permissible conditions might, for example, require double glazing, solid doors, insulation or a particular type of wall construction.[9]

9.2 ENTERTAINMENT

The Public Health Acts Amendment Act 1890 requires that houses, rooms, gardens or other places kept or used for public dancing, singing, music or other public entertainment of a like kind be licensed for such purposes.[10] Licences may be granted upon such terms and subject to such restrictions (including terms and restrictions relating to noise and vibration) as the District Court determines. Breach or disregard of a term or restriction is punishable by a penalty not exceeding £20 plus £5 per day for a continuing offence. The licence may also be revoked. The Public Dance Halls Act 1935 exercises somewhat similar control over public dancing with one major difference, viz. that any interested individual or any member of the Garda Siochana has a right to object to the granting or renewal of public dancing licences.[11] Objections raised often include objections to noise caused by dancing and associated activities. The Supreme Court has held in re. *Application of Quinn*[12] that one of the relevant matters which a District Justice has a general discretion to take into account under section 2(2) of the Public Dance Hall Act 1935, when considering an application for a licence, is the 'noisy conduct, disorderly behaviour and obscene language' of dance hall patrons.

Certain other social activities must also be authorised by the District or Circuit Courts. Clubs, for example, must be certified under the Registration of Clubs (Ireland) Act 1904. Section 5 of that Act provides as follows: 'The Court shall not consider any objection to the grant or renewal of a certificate unless it is taken on one or more of the following grounds'. Among the enumerated categories of grounds on which objection may be taken are that the club is 'habitually used for any unlawful purpose'. In *Comhaltas Ceolteoiri Eireann*[13] the High Court held that the habitual use of club premises in plain contravention of the Planning Acts would constitute an unlawful use of those premises justifying the District Court in refusing the grant or renewal of a certificate of registration.

9.3 CONTROLS OVER DESIGN AND CONSTRUCTION OF NOISE GENERATING PLANT AND EQUIPMENT

There are no specific statutory controls over the design and construction of noise generating plant and equipment. The Industrial Development Authority may require that plant and equipment be so designed as not to create excess noise as a condition for grant-aids. Under the Factories Act 1955, the Minister for Labour, if satisfied that any manufacture, machinery plant, equipment, appliance, process or description of manual labour is of such a nature as to cause risk of bodily injury to persons employed may, after consultation with the Minister for Health, make such special regulations as appear to him to be reasonably practicable and to meet the necessity of the case.[14] This power also exists in relation to work carried out on ships and to building operations. The Minister for Labour has made the Factory (Noise) Regulations 1975,[15] requiring the occupier of a factory within the meaning of the Factories Act 1955 to ensure that sound pressure levels which are likely to harm employees are kept at the lowest practicable level. Where sound pressure levels exceed 90 dBA, certain precautions must be taken. Similar requirements are contained in the Quarries (General) Regulations 1974.[16] The Safety in Industry Act 1980 requires occupiers of factories or specified premises to take appropriate steps to reduce sound levels to ensure that the hearing or health of their employees is not adversely affected. The Minister for Labour may make regulations requiring occupiers to carry out their duties in this respect.[17] These regulations are enforced by the Industrial Inspectorate attached to the Department of Labour.

Other statutory provisions relating to noise described in this section may operate indirectly as controls over the design and equipment of noise generating plant and equipment.

9.4 GENERAL RESTRICTION ON EMISSIONS OF NOISE AND VIBRATION

Section 51 of the Local Government (Planning and Development) Act 1963, as amended,[18] provides that it shall be an offence punishable on summary conviction by a fine not exceeding £50 plus £10 for each day the offence continues for any person:

(a) in any public place or in connection with any premises which adjoins any public place and to which the public are admitted, or

(b) upon any other premises, either:

(i) by operating, or causing or suffering to be operated any wireless, loudspeaker, television, gramophone, amplifier, or similar instrument, or any machine or other appliance, or

(ii) by any other means,

to make or cause to be made, any noise or vibration which is so loud, so continuous or so repeated or of such duration or pitch or at such times as to give reasonable cause for annoyance to persons in any premises in the neighbourhood or to persons lawfully using any public place.

Proceedings may not be taken by any person for an offence in respect of any premises referred to in (b) above unless the annoyance is continued after the expiration of 7 days from the date of a notice alleging annoyance, signed by not less than three persons residing or carrying on business within the area in which the noise or vibration is felt.[19] This means that there is no effective statutory remedy for preventing occasional or once-off noise and vibration. There may also be difficulties, especially in small, closely-knit communities, in getting three complainants to sign the notice alleging annoyance. A local authority is not empowered under the section to prosecute under section 51(1)(*b*), but it may prosecute under section 51(1)(*a*) since the nuisance involved here is essentially a public nuisance. There are saving clauses for noise caused by aircraft or by statutory undertakers in the exercise of powers conferred on them.[20] In proceedings brought in respect of noise or vibration caused in the course of a trade or business or in performing any statutory functions, it is a good defence for the defendant to prove that the best practicable means have been used for preventing and for counteracting the effect of the noise or vibration.[21] These exemptions from the scope of section 51(1) are unjustifiable and would hardly be necessary had planning authorities consistently ensured that suitable noise and vibration conditions were attached to permissions for developments. At present, it is very difficult to secure the abatement of noise and vibration nuisance where breaches of planning controls are not involved.

9.5 ENFORCEMENT AND INDIVIDUAL RIGHTS

Local authorities are primarily responsible for the enforcement of noise and vibration controls applicable under the Local Government (Planning and Development) Acts 1963 and 1976, although private individuals have increasingly participated in the enforcement of planning controls since the 1976 Act came into force. A number of applications have been made to the High Court under section 27 of the 1976 Act for what was essentially an order for the abatement of noise nuisance.[22] Many of these applications were brought by private individuals. This is as might be expected as noise and vibration nuisances are often very easily perceivable breaches of planning control.

There are no national statistics on the extent to which local authorities enforce section 51 of the 1963 Act, but the annual reports of Dublin Corporation for 1979 and 1980 indicate the extent to which noise problems occupy the attention of the Corporation's Environment Department.

	1978–9	1979–80
Number of complaints investigated	58	55
Number of complaints where action not possible (insufficient evidence, unfounded complaints, insufficient number of complaints)	44	43
Nuisance abated	14	12
Number of planning applications checked	105	155
Number of planning recommendations	77	114

Undue noise or vibration may also amount to a wrongful interference with a person's common law rights in which case an action may lie in trespass, nuisance, or negligence.[23] It is generally felt that noise nuisances are outside the rule in *Rylands* v. *Fletcher*.[24]

9.6 MOTOR VEHICLES

9.6.1 Implementation of EEC Directives

The European Communities (Motor Vehicle Type Approval) Regulations 1978[25] implement the provisions of certain EEC Directives which relate to the type approval of motor vehicles, trailers and components. The relevant Directives are Directives 70/157/EEC of 6 February 1970, as amended, relating to permissible sound levels and the exhaust systems of motor vehicles; the European Communities (Vehicle Type Approval) Regulations 1980,[26] which implement Directive 74/151/EEC of 4 March 1974 relating to certain parts and characteristics of wheeled agricultural or forestry tractors; and Directive 77/331/EEC of 29 March 1977, relating to the driver-perceived noise level of wheeled agricultural or forestry tractors. All of the above Directives fix permissible limits for the sound level of regulated vehicles and components and prescribe the equipment, conditions and methods for measuring this level. The enforcement of these Directives has already been described.[27]

9.6.2 National requirements

9.6.2.1 CONTROLS OVER DESIGN AND CONSTRUCTION

The Road Traffic (Construction Equipment and Use of Vehicles) Regulations 1963, as amended,[28] require that every vehicle be fitted with an audible warning device but, subject to certain exceptions, gongs, sirens, and other strident-toned devices are forbidden.[29] They also require that every vehicle be fitted with an exhaust silencer or other suitable device for reducing noise to a 'reasonable level'.[30]

9.6.2.2 CONTROLS OVER MAINTENANCE

The aforesaid Regulations require that every vehicle and parts and equipment thereof (which includes silencers and similar contrivances) be maintained in good and efficient working order. The alteration of a silencer in such a way as to increase noise is an offence.[31] All parts of a vehicle subject to severe vibration and all parts relevant to the control of a vehicle are required to be efficiently fastened so as to prevent their working or coming loose.[32]

9.6.2.3 CONTROLS OVER USE

Vehicles—whether mechanically propelled or not[33]—must not be used on public roads while there are attached to the vehicles a public address system incorporating a loudspeaker or similar device. Exceptions are made and there is provision for local authority licensing of such use of vehicles.[34] Vehicles may not be used in a public place so as to cause excessive noise which could have been avoided by the use of reasonable care.[35] Nor may audible warning devices be sounded between 23.00 hours and 7.00 hours in certain areas.[36] The use of any vehicle without an exhaust silencer or other similar contrivance is an offence.[37] Races on public roads are not permitted without the prior consent of the appropriate local authority.[38]

9.6.2.4 ENFORCEMENT

The above regulations are enforceable by the Garda Siochana only. There are no published statistics on the extent to which they are enforced. The penalty on conviction for breach of the regulations is £20 for a first offence and £50 for a second or subsequent offence.[39]

9.7 AIRCRAFT

Noise and vibration caused by aircraft are specifically exempted from the provisions of section 51 of the Local Government (Planning and Development) Act 1963.[40]

Ireland has accepted noise standards for aircraft agreed upon at the International Aviation Organisation meeting in Montreal in 1970. The Air Navigation (Noise Certificate and Limitation) Order 1976[41] gives effect to Amendment 2 to Annex 16 of the Chicago Convention and contains a number of controls over noise caused by subsonic aircraft. Noise abatement procedures must be followed and scheduled flights of jet aircraft are not permitted between 23.00 hours and 7.00 hours, but charter flights are allowed. Runway selection takes into account the need to avoid populated areas. These controls are enforceable by the Department of Transport.

There is no special legislation in force in Ireland at present concerning supersonic flights of civil and military aircraft. It is understood that such legislation has been drafted and can be introduced if necessary.

Restrictions on actions for trespass and nuisance caused by aircraft noise are contained in section 55 of the Air Navigation and Transport Act 1936, which provides that

> no action shall lie in respect of trespass or in respect of nuisance, by reason only of the flight of aircraft over any property at a height above the ground, which, having regard to wind, weather and all the circumstances of the case is reasonable, or the ordinary incidents of the flight, so long as the provisions of part II of this Act and any order made under the said part II and any regulations made by virtue of any such order are duly complied with.

Notes

1. See 9.6.
2. See 1.5.3.
3. See 2.1.1.
4. Local Government (Planning and Development) Act 1963, Third Schedule, Part II.
5. *Calculation of Road Traffic Noise* (1975), HMSO, London.
6. O'Cinneide, D., 'The Use of UK Methods for Traffic Noise Prediction in Ireland', *Irish Journal of Environmental Science* Vol. 1, pp. 78–79. Healy, J., 'Local Authority Monitoring Experience', Paper at Seminar of Noise (1980), National Board for Science and Technology.
7. Air Navigation and Transport Act 1950, s. 14.
8. Local Government (Planning and Development) Act 1976, s. 39(*c*).
9. Circular PD 103/1 of 30 July 1976.
10. Public Health Amendment Act 1890, s. 51.
11. Public Dance Halls Act 1935, s. 2(3).
12. In re. *Application of Quinn* [1974] I.R. 19.
13. In re. *Comhaltas Ceolteoiri Eireann*, H.C. 14.12.1977 (unreported).
14. Factories Act 1955, ss. 71, 83–89.
15. Factories (Noise) Regulations 1975 (S.I. No. 235 of 1975).
16. Quarries (General) Regulations 1974 (S.I. No. 146 of 1974).
17. Safety in Industry Act 1980, s. 13.
18. Local Government (Planning and Development) Act 1976, s. 40.
19. Local Government (Planning and Development) Act 1963, s. 51(3).
20. *Ibid.*, s. 51(5)(*a*) and (*b*).
21. *Ibid.*, s. 51(6).
22. See 2.4.8.
23. See 1.2.2.
24. Salmond, *The Law of Torts* (16th edn), p. 322.
25. S.I. No. 305 of 1978.
26. S.I. No. 41 of 1980.
27. See 3.9.1.

28. S.I. No. 190 of 1963.
29. Road Traffic (Construction Equipment and Use of Vehicles) Regulations 1963, art. 28.
30. *Ibid.*, art. 29.
31. *Ibid.*, art. 34(1).
32. *Ibid.*, art. 34(2).
33. Road Traffic (Construction Equipment and Use of Vehicles) (Amendment) Regulations 1965 (S.I. No. 79 of 1965).
34. Road Traffic (Construction Equipment and Use of Vehicles) Regulations 1963, art. 35.
35. *Ibid.*, art. 85(2).
36. *Ibid.*, art. 85(3).
37. *Ibid.*, art. 85(2).
38. *Ibid.*, art. 88.
39. Road Traffic Act 1961, ss. 11(4), 103; Road Traffic Act 1968, s. 8.
40. Local Government (Planning and Development) Act 1963, s. 51(5).
41. S.I. No. 250 of 1976.

10
Radioactive Substances and Nuclear Energy

Radioactive substances are used in hospitals, research institutions and industry. There is no nuclear power station in Ireland and plans for the erection of one in the 1980s have apparently been shelved. Responsibility for matters relating to nuclear energy and radioactive substances lies mainly with the Minister for Energy and the Nuclear Energy Board under the Nuclear Energy Act 1971. The Ministers for Labour and Health also enjoy limited control powers over the use of radioactive substances in factories and hospitals under the Factories Act 1955 and the Health Act 1956, respectively. The main source of law on radioactive substances is, however, the Nuclear Energy Act 1971. While the scope of sections 5 and 6 of this Act is extensive, the Act is defective in many ways. It does not deal with civil liability for damage caused by radioactive substances. The provisions in it on criminal liability are grossly inadequate. It provides for the regulation of matters of extraordinary importance by subordinate legislation. The powers of the Minister for Energy or the Nuclear Energy Board to revoke licences are limited to when they are of the opinion that licence conditions have been broken: there is no provision for revoking or amending licences in the interests of public health or safety. There are no special or detailed provisions dealing with the development of nuclear installations or the disposal of radioactive wastes.

There is no officially sanctioned depository for the storage, treatment or disposal of radioactive wastes in the country. There are no mandatory standards for levels of radioactivity in the environment.

193

10.1 CONTROLS BY LAND-USE PLANNING

Controls over radioactive substances may be exercised by conditions attached to planning permissions and approvals in cases where planning permission is necessary for new development.[1] In so far as can be discovered, the only instance where this has actually happened was in 1979 when Trinity College Dublin obtained permission for the development of storage facilities for radioactive wastes.[2]

10.2 CONTROLS OVER STORAGE, ACCUMULATION AND USE

Section 59 of the Health Act 1953, as amended,[3] specifically empowers the Minister for Health to make regulations 'for the control of the storage, exportation or other disposal of medical radioactive substances generally or of any particular radioactive substance'. This section has never been activated and most of the Minister for Health's functions in this respect have been assumed by the Nuclear Energy Board.

Under sections 6, 20 and 71 of the Factories Act 1955, the Minister for Labour has made the Factories Ionising Radiation (Sealed Sources) Regulations 1972,[4] and the Factories Ionising Radiations (Unsealed Radioactive Substances) Regulations 1972,[5] under which protective measures must be observed in factories to which the regulations apply.

The main source of law relating to radioactive substances is now the Nuclear Energy Act 1971, and the Nuclear Energy (General Control of Fissile Fuels, Radioactive Substances and Irradiating Apparatus) Order 1977.[6] Article 4 of the Order, *inter alia*, prohibits the custody and use of radioactive substances, radioactive devices, irradiating apparatus and radioactive waste products save in accordance with a licence issued by the Nuclear Energy Board as agent for the Minister for Energy. The Order is intended to complement controls exercisable by the Ministers for Health and Labour. It does not, however, apply to the medical uses of ionising radiation for the prevention, diagnosis and treatment of patients 'in order not to impinge on the doctor–patient relationship'.[7] Some apparatus and manufactured items may be exempted from licensing requirements under article 5 of the Order, provided they are of a type approved by the Nuclear Energy Board and comply with restrictions specified therein.

One of the particular functions of the Nuclear Energy Board under section 5(1) of the Nuclear Energy Act is to prepare draft safety codes and regulations dealing with fissile fuel or other radioactive substances or devices and irradiating apparatus, taking into account relevant standards recommended by international bodies dealing with nuclear energy.

The Board has not published any of these codes but it does exercise control over the use, accumulation and storage of radioactive substances, by conditions attached to licences which require the observance of recommended practices and procedures. Section 5(2) of the Nuclear Energy Act provides that the Minister for Energy may by order assign to the Board a number of functions including the making of arrangements to ensure the safe custody of fissile fuel, the safe operation of radioactive devices and the making of arrangements to ensure compliance with any safety codes established or regulations made under any of the three Acts mentioned in this section. The Minister has not made any order under section 5(2).

10.3 CONTROLS OVER PACKAGING AND TRANSPORT

10.3.1 General

Controls over the packaging and transport of radioactive substances are exercised by virtue of conditions attached to licences granted by the Nuclear Energy Board. In licensing the transportation of radioactive substances the Board requires that International Atomic Energy Regulations be observed.[8] When necessary the Board imposes additional specific conditions appropriate to the particular licensee. In cases where large consignments of radioactive materials are involved, the Board will make special arrangements for their transport in Ireland, including the provision of an appropriate escort.

Under section 5 of the Nuclear Energy Act 1971, the Minister for Energy may, by order, assign to the Nuclear Energy Board the function of making arrangements for the transportation of fissile fuel or such other radioactive substances or devices as he specifies in the order. Furthermore, the transportation of fissile fuel, or of such other radioactive substances or devices or irradiating apparatus, including radioactive waste products, as are specified is a matter which may be regulated directly by Ministerial order under section 6 of the Act. The Minister has not made any of these orders.

10.3.2 Transport by rail

The transport of 'dangerous goods' by Coras Iompair Eireann (the National Transport Company) is regulated under section 60 of the Transport Act 1960, which deals with the transport of dangerous goods by rail. The section provides that nothing is to impose any obligation on CIE to accept dangerous goods by rail but that if such goods are accepted, they are to be conveyed subject to such by-laws, regulations and conditions as CIE think fit in regard to the conveyance and storage thereof; that the owner or consignor must indemnify CIE for all loss or damage which may result to it or to which it may become liable owing to non-compliance with the said by-laws, regulations or conditions; and that full compensation must be paid to CIE or its servant for injuries or damage save when the injury or damage is due to the wilful misconduct of CIE servants. CIE has not made specific provisions relating to the carriage of radioactive substances in its legislation.

10.3.3 Transport by sea

The packaging and transport of radioactive substances by sea is subject to two kinds of controls, viz. controls applicable to the transport of 'dangerous goods' and controls enacted with specific reference to radioactive materials.

The Merchant Shipping Act 1894, as adapted and applied in Ireland, contains provisions on the carriage of dangerous goods. Under section 38(4) of the Merchant Shipping (Safety Convention) Act 1952 and the Merchant Shipping (Dangerous Goods) Rules 1967,[9] radioactive materials come within the statutory definition of dangerous goods and are subject to the requirements of section 446 of the Merchant Shipping Act 1894 and the Merchant Shipping (Dangerous Goods) Rules 1967 (the Rules implement the provisions of the International Convention for the Safety of Life at Sea 1960, relating to the carriage of dangerous goods).

10.3.4 Transport by air

Effect has been given to the Chicago Convention on International Air Aviation and amendments thereto in the Air Navigation and Transport Act 1946. Under section 6 of the 1946 Act the Minister for Transport has made the Air Navigation (Carriage of Munitions of War, Weapons and Dangerous Goods) Order 1973[10] which operates to control the carriage of radioactive substances by air.

10.3.5 Sending by post

The sending of dangerous goods by post is prohibited by section 63 of the Post Office Act 1908. The Post Office Regulations set out in the Post Office Guide[11] state that radioactive materials with any significant alpha, beta, gamma or neutron radiation may not normally be transmitted by post. In exceptional circumstances the transmission of small quantities is permitted on condition that they are packed in accordance with the regulations and provided that when made up for postage the radiation measured at the outside of the packet does not exceed 10 milliroentgens per 24 hours. Containers must be submitted for inspection and approved for use. The Universal Postal Union, of which Ireland is a member, permits the transmission by post of radioactive materials whose contents and make-up comply with the regulations of the International Atomic Agency and with the provisions of the detailed regulations of the Universal Postal Union.

10.4 CONTROL OVER RADIOACTIVE WASTE DISPOSAL[12]

Statutory responsibility for controlling the disposal of radioactive wastes lies exclusively with the Nuclear Energy Board, the Minister for Health's functions in this regard having been terminated by section 24(c) of the Nuclear Energy Act 1971. The Nuclear Energy Board has power to regulate the disposal of radioactive wastes by conditions attached to licences. Limits have been laid down to control the discharge of small quantities of low level radioactive effluents to the environment. The Board also exercises control over methods to process radioactive wastes so that suitable management systems are used. Hospitals and laboratories usually incinerate slightly contaminated combustible refuse produced daily. Interim storage facilities for solid radioactive wastes are provided by a Dublin hospital specialising in cancer treatment. Spent radioactive sources are returned to their suppliers abroad. There are no officially approved facilities for the disposal of radioactive wastes.

10.5 LEGAL STANDARDS, OBJECTIVES AND GUIDELINES FOR LEVELS OF RADIOACTIVITY IN THE ENVIRONMENT

There are no legally binding standards, objectives or guidelines for levels of radioactivity in the environment. The Nuclear Energy Board in preparing licence conditions takes note of recommendations and guidelines issued by such international organisations as the International Commission on Radiological Protection, the European Communities, the International Atomic Energy Agency, the Nuclear Energy Agency, the World Health Organisation and others.[13]

10.6 MONITORING AND ENFORCEMENT

The provisions of the Factories Ionising Radiations (Sealed Sources) Regulations 1972 and the Factories Ionising Radiations (Unsealed Radioactive Substances) Regulations 1972 are enforced by the Industrial Inspectorate of the Department of Labour or, where appropriate, by officers of the Nuclear Energy Board.[14] The Board operates a monitoring and inspection system to ensure that conditions attached to licences are observed and safety standards maintained. Over 150 licences were in operation in 1979 covering principally medical, research and industrial users of ionising radiation.[15] A National Radiation Monitoring Service is operated by the Nuclear Energy Board. In 1979 the service issued over 40,000 personal dosimeters to persons dealing with radioactive substances.[16] It also monitors radioactivity in the Irish Sea. The Meteorological Service of the Department of Transport carries out regular measurements of the radioactivity of precipitation, settled dust and airborne particles in a number of selected locations throughout the country.

Contravention of the provisions of section 6 of the Nuclear Energy Act 1971 (establishing a licensing system) is an offence for which the maximum penalty is a fine not exceeding £500 and/or 5 years' imprisonment.[17] Non-compliance with the provisions of the Merchant Shipping Act 1894, relating to the carriage of dangerous goods by sea, is punishable by a maximum fine of £500[18] and, perhaps, forfeiture of the goods.[19] The maximum fine for breach of the Merchant Shipping (Dangerous Goods) Rules 1967 is £300. Contravention of regulations made under section 71 of the Factories Act 1955 is punishable by a fine not exceeding £20 and £5 for each day the offence is continued.[20] In so far as can be discovered, no prosecutions have been brought in respect of breach of any of the above-mentioned Acts or regulations.

10.7 INDIVIDUAL RIGHTS

There are no special provisions in legislation dealing with radioactive substances on individual rights.

Notes

1. See Chapter 2.
2. See *Annual Report of the Nuclear Energy Board* (1979).
3. Health Act 1953, s. 59, as amended by Nuclear Energy Act 1971, s. 24.
4. S.I. No. 17 of 1972.
5. S.I. No. 249 of 1972.
6. S.I. No. 166 of 1977.
7. *Nuclear Energy Board News*, No. 2, May 1979.
8. *Ibid.*
9. S.I. No. 105 of 1967.
10. S.I. No. 224 of 1973.
11. *Post Office Regulations*, Vol. 1, p. 48.
12. *Ibid.*
13. *Ibid.*
14. Factories Act 1955, s. 93, and Nuclear Energy Act 1971, s. 25.
15. *Annual Report of the Nuclear Energy Board* (1979).
16. *Ibid.*
17. Nuclear Energy Act 1971, s. 6(4).
18. Merchant Shipping Act 1894, s. 447.
19. *Ibid.*, s. 449.
20. Factories Act 1955, s. 101.

11
Products

There is, as the Inter-departmental Committee on Pollution Control concluded, 'no overall system for the control of environmentally harmful chemicals and chemical products used in industry, commerce and domestically'.[1] There are, however, a number of statutes under which environmentally harmful substances in certain agrichemicals are subject to controls administered by the Departments of Agriculture and Health. The necessity of complying with the requirements of EEC Directives has also resulted in the enactment under the European Communities Act 1972 of controls over various other chemical products.

Restrictions on the use of chemical products used in agriculture is, in practice, almost solely by restriction on distribution. This is because manufacturers regard it as essential from a marketing point of view to get the approval of the Department of Agriculture for products which they wish to sell for agricultural use. The Department consults with the research staff of the Agricultural Institute before granting approval for any newly marketed chemical product but there is no national entity for approving the ecotoxicity of chemical products.

Advice on the use of chemicals in agriculture is provided to farmers by the Department of Agriculture, the Agricultural Advisory Services and the media.

11.1 FERTILIZERS, FEEDING STUFFS AND MINERAL MIXTURES

Control over chemical substances in these products is exercised by the Minister for Agriculture under the Fertilizers Feeding Stuffs and Mineral Mixtures Act 1955, the European Communities Act 1972, and regulations made under both Acts.

Matters relating to chemical substances in non-EEC fertilizers are regulated under the 1955 Act; the Fertilizers Feeding Stuffs and Mineral Mixtures Regulations 1957;[2] the Marketing of non-EEC Fertilizers Regulations 1978[3] and 1979,[4] and the Fertilizers Feeding Stuffs and Mineral Mixtures (Methods of Analysis) Regulations 1978[5] and 1980[6] made thereunder; EEC Council Directive 76/116/EEC and EEC Commission Directive 77/535/EEC on fertilizers, implemented by the European Communities (Marketing of Fertilizers) Regulations 1978[7] and 1979[8] and the European Communities (Sampling and Analysis of Fertilizers Regulations) 1978[9] and 1979[10] made under the European Communities Act 1972. Quality, packaging and labelling standards for straight and compound NPK solid fertilizers are prescribed under the above regulations as well as standardised methods of sampling and analysis of fertilizers.

Chemical substances in animal feeding stuffs are regulated under the Fertilizers Feeding Stuffs and Mineral Mixtures Act 1955, the Fertilizers Feeding Stuffs and Mineral Mixtures Regulations 1957,[11] the Fertilizers Feeding Stuffs and Mineral Mixtures (Methods of Analysis) Regulations 1978,[12] and the Fertilizers Feeding Stuffs and Mineral Mixtures (Methods of Sampling) and Fertilizers Feeding Stuffs and Mineral Mixtures (Methods of Analysis) (Amendment) Regulations 1980,[13] made under the 1955 Act. EEC Directives 70/524/EEC as amended from time to time and EEC Council Directive 75/296/EEC governing the use of additives in feeding stuffs have been implemented by the European Communities (Feeding Stuffs) (Additives) Regulations 1974[14] and 1979.[15] EEC Directives 74/63/EEC, 76/14/EEC and 76/934/EEC fixing maximum permissible levels of certain undesirable substances and products in animal feeding stuffs have been implemented in the European Communities (Feeding Stuffs) (Tolerance of Undesirable Substances and Products) Regulations 1977.[16] Standardised methods by which sampling and analysis of animal feeding stuffs are to be carried out for the purposes of the above-mentioned Tolerances of Undesirable Substances and Products and the Additives Regulations are prescribed in the European Communities (Feeding Stuffs) (Methods of Analysis) Regulations 1978[17] and the European Communities (Feeding Stuffs) (Methods of Analysis (Amendment) and Methods of Sampling) Regulations 1980.[18]

Manufacturers of feeding stuffs and mineral mixtures must be licensed by the Minister for Agriculture under section 5 of the Fertilizers Feeding Stuffs and Mineral Mixtures Act 1955. In December 1978 there were 190 licensed manufacturers of compound feeding stuffs and 18 licensed manufacturers of mineral mixtures.[19] The Minister has power to attach, add to or vary conditions attached to licences or to revoke a licence whenever he considers it proper to do so.[20]

Under section 6 of the 1955 Act the Minister for Agriculture may make regulations, *inter alia*, limiting or regulating the use of any specified article in the manufacture for sale of fertilizers, feeding stuffs or mineral mixtures, and prescribing standards and definitions for these products. The Minister has exercised his powers under section 6 to prohibit the use of arsenic and arsenical compounds in feeding stuffs and mineral mixtures[21] and to prohibit the use of specified antibiotics in the manufacture for sale of compound feeding stuffs.[22]

The inspectorate of the Department of Agriculture oversees the implementation of the above Acts and regulations. The penalty on summary conviction for breach of a regulation made under the 1955 Act is £25.[23] Breaches of regulations made under the European Communities Act 1972 are punishable on summary conviction by a maximum fine which rarely exceeds £200 and/or 6 months' imprisonment.

11.2 PESTICIDES

The only legislative controls over pesticides are contained in the European Communities (Pesticide Residues) (Fruit and Vegetables) Regulations 1979, as amended in 1981,[24] made by the Minister for Agriculture under the European Communities Act 1972. These regulations prohibit the putting into circulation of any fruit or vegetable to which the regulations apply if it contains the residue of a pesticide specified in the regulations in a quantity greater than the maximum allowed for that pesticide. Other controls over pesticides are of a non-statutory nature and operate solely by restrictions on distribution. The Department of Agriculture is preparing a pesticide registration scheme with the aid of an Advisory Committee. This scheme, when implemented, will prohibit the use of certain pesticides based on considerations which will include residue effects, toxic properties, efficacy considerations and effects on the environment and wild-life.[25]

11.3 VETERINARY PRODUCTS

A voluntary, non-statutory veterinary products control scheme is administered by the Department of Agriculture to ensure the proper use and control of veterinary products. It is anticipated that this will become mandatory under EEC law in the near future.[26] The Minister for Agriculture has power under section 15 of the Poisons Act 1961 to make regulations as to the use of poisons for agricultural or veterinary purposes but he has not yet availed of these powers.

11.4 ANIMAL REMEDIES

Under section 7 of the Animal Remedies Act 1956, the Minister for Agriculture is empowered to control the manufacturing, preparation, packaging, importation or sale of any animal remedies. Section 5 of the Act requires that the composition of animal remedies, the proportion of ingredients therein, the commercial or scientific name of the remedy and the specific remedial property or properties claimed for the remedy be disclosed on containers of animal remedies and in advertisements. Containers must also indicate the name and address of the manufacturer, packer, distributor or importer, as appropriate. The Minister has made two sets of regulations under the Act prohibiting the sale, except under licence, of any anti-abortion vaccine for use in the vaccination of cattle against brucellosis[27] and for the sale of chloramphenicol or its salts or derivatives or any preparation containing such substances.[28]

Section 35 of the Diseases of Animals Act 1966 and regulations made thereunder provide that only sheep dips approved by the Minister for Agriculture may be used for dipping sheep for the prevention or treatment of sheep scab.[29] The Minister for Agriculture has withdrawn his approval for the use of dieldrin and aldrin pesticides.

11.5 WEEDKILLERS

Certain weedkillers may be controlled by regulations made under the Poisons Act 1961. The Poisons Act (Paraquat) Regulations 1975,[30] passed more to limit accessibility to the favourite Irish suicide potion than for environmental reasons, prohibit the retail of paraquat by persons other than pharmaceutical chemists or persons entitled to sell other poisons, specify packaging and labelling precautions and require sellers to keep records of all sales. There are no other statutory controls over the sale of weedkillers.

11.6 MISCELLANEOUS

The European Communities (Detergents No. 2) Regulations 1975[31] and the European Communities (Aerosol Dispensers) Regulations 1977,[32] made under the European Communities Act 1972, implement Council Directive 73/404/EEC and Council Directive 73/324/EEC respectively. The Detergents Regulations prohibit the placing on the market and use of detergents where the average level of biodegradibility of certain essential constituents is less than 90%. The Aerosol Dispensers Regulations lay down conditions and standards governing the manufacture, filling, marketing and labelling of certain aerosol dispensers. The European Communities (Prohibition of Certain Active Substances in Plant Protection Products) Regulations 1981[33] implement Council Directive 79/117/ EEC and provide that plant protection products containing certain active substances (certain mercury compounds and persistent organo-chlorine compounds) may not be placed on the market or used except as permitted in the Regulations. Guidelines on precautions to be observed in the use of toxic chemicals in agriculture and horticulture are circulated to farmers by the Department of Agriculture.[34]

Notes

1. *Report on Pollution Control*, p. 18.
2. S.I. No. 264 of 1957.
3. S.I. No. 248 of 1978.
4. S.I. No. 410 of 1979.
5. S.I. No. 249 of 1978.
6. S.I. No. 13 of 1980.
7. S.I. No. 13 of 1973.
8. S.I. No. 411 of 1979.
9. S.I. No. 12 of 1978.
10. S.I. No. 409 of 1979.
11. S.I. No. 264 of 1957.
12. S.I. No. 249 of 1978.
13. S.I. No. 13 of 1980.
14. S.I. No. 302 of 1974.
15. S.I. No. 6 of 1979.
16. S.I. No. 246 of 1977.
17. S.I. No. 250 of 1978.
18. S.I. No. 14 of 1980.
19. *Annual Report of the Department of Agriculture* (1978).
20. Fertilizers Feeding Stuffs and Mineral Mixtures Act 1955, s. 5(3).
21. Animal and Poultry Feeding Stuffs and Mineral Mixtures (Control of Arsenic) Regulations 1972 (S.I. No. 124 of 1972).
22. Animal and Poultry Compound Feeding Stuffs (Control of Antibiotics) Regulations 1972 (S.I. No. 335 of 1972).
23. Fertilizers Feeding Stuffs and Mineral Mixtures Act 1955, s. 6(3).
24. S.I. No. 183 of 1979; S.I. No. 164 of 1981.
25. *Report on Pollution Control*, p. 17.
26. *Ibid.*
27. Animal Remedies (Control of Certain Anti-Abortion Vaccines) Regulations 1965 (S.I. No. 112 of 1965).
28. Animal Remedies (Control of Chloramphenicol) Regulations 1975 (S.I. No. 10 of 1975).
29. Sheep Dipping Orders 1965, 1966, 1976, 1977 (S.I. Nos. 105 of 1977, 98 of 1965, 107 of 1966, and 158 of 1978).
30. S.I. No. 146 of 1975.

31. S.I. No. 107 of 1975.
32. S.I. No. 144 of 1977.
33. S.I. No. 320 of 1981.
34. *Guidelines in the Use of Toxic Chemicals in Agriculture and Horticulture*, Farm Bulletin, November 1975.

12
Environmental Impact Studies

The Irish requirement for environmental impact statements—called studies—is contained in section 39(*a*) of the Local Government (Planning and Development) Act 1976, which amends section 25(2) of the Local Government (Planning and Development) Act 1963, and in article 20 of the Local Government (Planning and Development) Regulations 1977. Section 25(2)(*cc*) of the 1963 Act empowers the Minister for the Environment to make regulations providing for the furnishing to planning authorities in cases where development to which a planning application relates 'of a written study of what, if any, effect the proposed development, if carried out, would have on the environment relative to the place where that development is to take place', but the obligation is to apply only 'in cases where the development to which the application relates will, in the opinion of the relevant planning authority, cost more than the amount specified in the regulations'. Article 28 of the regulations provides:

(1) An application to a planning authority for a permission for any development to which this article applies shall, notwithstanding the provisions of article 18 or, in the case of an outline application, article 19, be accompanied by two copies of a written study of what, if any, effect the proposed development, if carried out, would have on the environment relative to the place where the development is to take place.

(2) Where a planning authority receives an application for a permission for a development which, in their opinion, is a development to which this article applies and the application is not accompanied by a written study as required by sub-article (1), they may, in addition to their powers, under article 26, require the applicant to submit such written study.

(3) This article applies to any development:

(a) for the purposes of any trade or industry (including mining) comprising any works, apparatus or plant used for any process which would result in the emission of noise, vibration, smell, fumes, smoke, soot, ash, dust or grit or the discharge of any liquid or other effluent (whether treated or untreated) either with or without particles of matter in suspension therein and,

(b) the cost of which, including all fixed assets as defined in section 2 of the Industrial Development Act 1969 (No. 32 of 1969), may reasonably be expected to be five million pounds or more.

The obligation to make an environmental impact study (EIS) appears to apply solely to developments in respect of which a planning application is made. This means that exempted development as defined in section 4 of the 1963 Act and in the Local Government (Planning and Development) Regulations 1977, and development by State Authorities as defined by section 84 of the 1963 Act, are not subject to EIS requirements because planning applications are not required in respect of these developments.[1] These exemptions from the provisions of section 25(2)(cc) are extensive and significant. Section 4 of the 1963 Act exempts, *inter alia*, developments by local authorities in their own areas and developments consisting of the use of any land for the purpose of agriculture or forestry. Among the more important exemptions (from the environmental point of view) contained in the regulations are developments by harbour authorities, much development consisting of agricultural buildings, and the storage in industrial buildings of wastes (including toxic wastes). In section 84 of the 1963 Act, 'State Authorities' means members of the Government, the Commissioners of Public Works and the Land Commission—bodies in whose names a significant amount of development is carried out. In addition, since offshore development is not regulated under the Planning Acts, there is no obligation to submit an EIS in respect of proposals for such development. In general, therefore, it may be said that an EIS will not be required for most public projects and that many public authorities in Ireland enjoy an immunity from environmental protection controls and from public scrutiny which is not available to their counterparts in the private sector. The Irish provisions are therefore applicable mainly to private industrial development in its narrow sense. To some extent, therefore, the 1976 legislation on EISs merely institutionalises and makes obligatory the existing administrative practice whereby IDA-sponsored projects are subjected to a limited form of environmental assessment by the IIRS and An Foras Forbartha.[2]

Environmental impact studies are required in respect of development

207

which in the opinion of the relevant planning authority may be expected to cost £5 million or more. This delineation of development subject to EIS requirements by reference to its cost is a crude attempt to distinguish between major and minor developments, and it makes little sense from an environmental point of view. The cost of a proposed development is only one of the relevant factors in considering whether it could have major consequences for the environment. The fact that the relevant planning authority—if they take the obligation seriously—must consider whether or not the proposed costs may exceed £5 million presumably involves them in financial investigations which are time-consuming and of little benefit to the environment. In any case, a developer could fail to satisfy the cost criterion by fragmentation of his project.

There may be proposed developments which do not clearly and unambiguously satisfy the criteria in article 28(3). In such cases article 28(2) gives planning authorities a discretion as to whether or not they will require an EIS. The possibility of such cases occurring must necessarily be limited as whether or not a development will result in the emission of pollutants and cost more than £5 million is usually a question of fact. But there may, as article 28(2) envisages, be borderline cases. In such cases a planning authority is bound to *consider* whether an EIS is required before coming to a positive or negative conclusion on the matter.[3]

The possibility of having a decision set aside might appear to be slightly academic in view of section 82(3) of the Local Government (Planning and Development) Act 1963, as amended,[4] which provides that:

A person shall not by prohibition, certiorari or in any other legal proceedings whatever question the validity of

(a) a decision of a planning authority on an application or approval under Part IV of the Principal Act.

(b) a decision of the Board on any appeal or on any reference.

(c) a decision of the Minister on appeal, unless the proceedings are instituted within the period of two months commencing on the date on which the decision is given.

But, in practice, the institution of proceedings to question the validity of a decision could occasion unacceptable delays for a developer and it is by no means certain that the prohibition in section 82(3A) will always be upheld by the courts.[5]

The legal requirements as to EISs do not specify what should be contained in such a study other than the 'effect which the proposed development, if carried out, would have on the environment relating to the place where the development is to take place'. Thus the EIS may be confined to the effect of the proposed development on the place where it is to be situated and not, presumably, on other areas. But what does 'effect' mean? Does it mean good effects and/or bad effects, all of the effects or only some of them, major effects or minor effects? As yet there are no official answers to these questions.

Notes

1. See 2.4.1.
2. See 1.5.1, 1.5.2, 1.5.3.
3. *Anns* v. *London Borough of Merton* [1976] 2 All E.R. 492.
4. Local Government (Planning and Development) Act 1976, s.42.
5. Gravells, N.P., 'Time limit clauses and judicial review: the relevance of context' (1978), 41 Modern Law Review, 383; 'Time limit clauses and judicial review — some second thoughts' (1980), 43 Modern Law Review, 173. For examples of judicial ingenuity in circumventing section 82(3), see *The State (Pine Valley Developments Ltd.)* v. *Dublin County Council*, H.C. 27.5.1981 (unreported) and *Freeney* v. *Bray U.D.C.*, H.C. 16.7.1981 (unreported).

Classified Index*

The Constitution, Public Authorities, Special Interest Groups and Individuals

The national constitution	1.1
Sources of laws governing pollution control and remedies for damage caused by pollution	1.2
Government departments and agencies with supervisory, administrative or executive powers of pollution control	1.3
National, regional and local public authorities with powers of pollution control	1.4
Independent advisory bodies with rights or duties under pollution control legislation	1.5
Special interest groups representing those who may be liable for pollution, or those concerned to prevent or reduce pollution	1.6
Standing to sue (*locus standi*) in legal proceedings for pollution	2.4.5

Air

Stationary Sources

Control by land use planning	2, 3.5
Controls over plant and processes (including raw materials, e.g. fuels)	3.1, 3.2.1, 3.3, 3.6, 3.7
Controls over treatment before discharge, and over manner of discharge (e.g. height of chimney)	3.2.1, 3.5
Limits on emissions	3.2.1, 3.3, 3.4.2
Monitoring to be done by discharger	3.3
Enforcement, including monitoring and surveillance by or on behalf of the control authority	3.1, 3.2.3, 3.4.4, 3.12

*References are to section numbers.

211

Controls over Products

Environmental Impact Assessment

Controls over Products

Environmental Impact Assessment